THE REGIONAL DYNAMICS OF LANGUAGE DIFFERENTIATION IN BELGIUM

A STUDY IN CULTURAL-POLITICAL GEOGRAPHY

by

ALEXANDER B. MURPHY

UNIVERSITY OF OREGON

UNIVERSITY OF CHICAGO

GEOGRAPHY RESEARCH PAPER NO. 227

1988

Copyright 1988 by
The Committee on Geographical Studies
The University of Chicago
Chicago, Illinois

Library of Congress Cataloging-in-Publication Data
Murphy, Alexander Bailey, 1954 -
  The regional dynamics of language differentiation in Belgium: A study in cultural-political geography.

  (Geography Research paper ; no. 227)
  Bibliography: p. 205.
  Includes index.
   1. Belgium--Historical geography. 2. Language policy--Belgium. 3. French language--Political aspects--Belgium. 4. Dutch language--Political aspects--Belgium. 5. Belgium--History--1830-1914. 4. Belgium--History--1914-   . I. Title. II. Series: Research Paper (University of Chicago. Committee on Geographical Studies) ; no. 227.
DH430.M87 1988   949.3   88-11853
ISBN 0-89065-132-9 (pbk.)

Geography Research Papers are available from:
The University of Chicago
Committee on Geographical Studies
5828 South University Avenue
Chicago, Illinois 60637-1583
Price: $12.00; $9.00 series subscription

To

the memory of my father

RICHARD ERNEST MURPHY

who gave me the world

CONTENTS

LIST OF FIGURES — vii

LIST OF TABLES — ix

ACKNOWLEDGMENTS — xi

NOTE ON PLACE NAMES — xiii

Chapter

1. INTRODUCTION — 1

2. ETHNICITY, REGIONALISM, AND THE SUBSTATE NATIONALIST QUESTION — 13

    Culture, Ethnicity, and Nationalism
    Perspectives on the Rise of Substate Nationalism
    Territoriality and Regionalism
    Perspectives on Linguistic Regionalism in Belgium

3. LANGUAGE AND REGIONALISM IN THE EMERGING BELGIAN STATE — 41

    Historical Introduction
    Language and Regionalism in the Newly Created Belgian State

4. AWAKENING ETHNOLINGUISTIC IDENTITY IN THE NINETEENTH CENTURY — 57

    Characteristics of the Newly Independent Belgian State
    The Rise of the Flemish Movement
    Legislative Developments in the Late Nineteenth Century
    Walloon Reactions
    An Overview of Linguistic Regionalism in the Nineteenth Century

5. THE RISE OF ETHNOREGIONALISM IN THE EARLY
   TWENTIETH CENTURY   91

   Language, Ethnicity, and Regionalism prior to
      World War I
   World War I and the Idea of Flemish Political Autonomy
   Postwar Reactions
   The Push for Territorial Unilingualism
   The Implications of Territorial Unilingualism
   World War II and Its Immediate Aftermath
   Summary

6. THE RESTRUCTURING OF BELGIUM ALONG
   LANGUAGE LINES   125

   The Growing Salience of Regional Divisions in the 1950s
   The Language Laws of the Early 1960s
   The End of the Unitary State
   The Status of the Language Regions in Belgium since 1970
   Summary

7. THE CHANGING CONTEXT OF GROUP IDENTITY
   AND INTERACTION   153

   Changing Regional Arrangements and Interaction Patterns
      Linguistic considerations
      Demographic factors
      Urban issues
      Changing political and governmental arrangements
      Economic dichotomization
      Social and cultural divisions
      Landscape factors

   Evolving Attitudes and Issues
      The promotion of collective ethnoregional identity
      The question of mutual understanding
      The focus on territorial issues
      The importance of interregional comparisons
      The emphasis on regional problems and possibilities

8. CONCLUSION   189

CHRONOLOGY   197

BIBLIOGRAPHY   205

INDEX   245

LIST OF FIGURES

1.	Language regions of Northwest Continental Europe	2
2.	Hapsburg provinces in area of modern Belgium - 1549	44
3.	Simplified relief regions of Belgium	54
4.	Simplified economic regions of Belgium	62
5.	Percentage of the work force engaged in agriculture in Belgium in 1846	81
6.	Percentage of illiterate applicants for military service in Belgium in 1867	85
7.	Indicators of strength of traditional Catholic values in Belgium in 1890	88
8.	Language regions of Belgium	133
9.	Communes of the Brussels Metropolitan Area	137
10.	Legislative institutions in Belgium	150

LIST OF TABLES

1.	Languages Most Frequently Spoken in Belgium, 1846-1947	5
2.	Population by Language Region, 1970-1984	6
3.	Percentage of Work Force in Agriculture, 1846	82
4.	Percentage of Work Force in Agriculture, 1856	83
5.	Percentage of Population in Industry, 1856	83
6.	Percentage of Illiterate Applicants for Military Service, 1867	84
7.	Percentage of Marriages Involving Divorcees, 1890	86
8.	Number of Inhabitants per Priest, 1890	86
9.	Migration of Francophones Listed in *Qui est qui en Belgique francophone*	160
10.	Migration of Dutch Speakers Listed in *Wie is Wie in Vlaanderen*	160

ACKNOWLEDGMENTS

I am deeply indebted to Professor Marvin W. Mikesell of the University of Chicago for his encouragement, advice, and support throughout the preparation of this study. From the inception of the problem to the completion of the final manuscript, he has always been an available and sympathetic listener, a thoughtful critic, and a friendly adviser. I am also grateful to Professors Michael P. Conzen, Chauncy D. Harris, and Norton S. Ginsburg for their counsel and careful reading of drafts of the manuscript. Each contributed fundamentally to my study with many helpful comments and suggestions.

Much of the primary research for this study was supported by a Fulbright-Hays grant to Belgium in 1985-1986, which I acknowledge with gratitude. I am also grateful to the Belgium-America Educational Foundation for an honorary fellowship while I was in Belgium. I appreciate the assistance of the Brussels offices of both organizations. An unexpected advantage of working under a Fulbright grant was to be brought into contact with other American scholars working on Belgium. Patricia Hilden, with whom I shared a table at the national library in Brussels, was an excellent listener, the supplier of countless provocative ideas, and a companion in the face of triumph and adversity. Janet Polasky has an extensive knowledge of Belgium past and present, and was always ready to share her experience and insights.

During my stay in Belgium, I benefited greatly from the assistance provided by many institutions and individuals. The Centrum voor de Studie van de Pluriculturelle Maatschappij/Centre pour l'Etude de la Société Pluriculturelle provided me with a base of operations from which to conduct my research. The Center's director, Mona Grinwis, and staff members Lieve Suenaert and Pierre Verdoodt, gave me access to the Center's holdings and facilities, put me into contact with people who shared my research interests or were knowledgeable about particular factual issues, and included me in many of the Center's activities. I am also grateful to the staff of the Bibliothèque Royale Albert 1er for providing me with considerable assistance and a place to work for many months. Particular thanks go to Lisette Danckaert, head of the map collection, and Eddy Bruynseels.

In addition, I was very fortunate to meet many Belgians who expressed interest in my study and who shared their time and insights. Piet Van de Craen and Anne Osterreith were most helpful and congenial. Piet Van de Crean deserves special thanks for carefully reviewing the manuscript and making a number of helpful suggestions. A number of individuals I met through my wife's association with the law firm of DeBandt, van Hecke & Lagae, helped me to gain a better understanding of their country through their openness and friendliness. I owe particular thanks to Guido and Marlene DeWit, Catherine Therasse, Philippe Latour, and Patrick Kelly. Catherine Therasse also provided great help in tracking down citations for me in Belgium after I returned to the United States.

I am grateful to the geography department at the University of Chicago for providing a stimulating and congenial environment and a pleasant place to work during the beginning stages of my research and as I drafted the manuscript. Many friends in the department helped along the way, including Jane Benson, Elizabeth Brooks, Michael Childs, Shaul Cohen, Chad Emmett, Peter Fendrick, Pavel Kraus, Mary McNally, Ranjana Patnaik, and Gloria Wolpert. Alexis Papadopoulos, who is also among this group, deserves special thanks for his assistance with the maps and figures. Thanks are also due to Susan Alitto for overseeing the preparation of the manuscript for publication and to Reine Mikesell for reviewing the French-language entries in the manuscript.

Finally I would like to express my deep appreciation to my family. My father, Richard E. Murphy, shared with me his passion for geography. My sister, Caroline K. Murphy, read the entire manuscript and made numerous helpful suggestions. My brother, R. Taggart Murphy, has been a continuing source of support and intellectual stimulation. My mother, Esther B. Murphy, provided encouragement and help throughout the research and writing. My greatest debt of gratitude is to my wife, Susan N. Gary, who not only made this possible, but made the process enjoyable and worthwhile.

NOTE ON PLACE NAMES

The language in which place names are rendered can be a politically sensitive issue in Belgium, and hence, the approach adopted here requires some explanation. Terms for place names in Belgium are generally rendered in the language of the region to which the place presently belongs by virtue of the Belgian law of November 8, 1962, dividing the country into language regions. An exception is made for those toponyms for which there is a widely accepted English equivalent. These are Brussels (Bruxelles/Brussel), Flanders (Vlaanderen), Antwerp (Antwerpen), Ghent (Gent), Bruges (Brugge), and Fourons (Voer). For purposes of clarity, the name for a place rendered in one of the major national languages is occasionally followed by its equivalent in the other major national language in parentheses. A particular problem arises in connection with references to the university in Leuven (Louvain). Although Leuven is within the Dutch-language region, the primary language of the university from the 1830s until the last few decades was French, and the institution was widely known during this period as the Université Catholique de Louvain or, in English, the Catholic University of Louvain. This study adopts this English appelation to refer to the university before the departure of the French-language section for Louvain-la-Neuve in the late 1960s. In keeping with the general approach to the rendering of place names outlined above, however, the city in which the university is located is always referred to as Leuven.

Chapter 1

INTRODUCTION

A compelling reality of contemporary Belgium is the pervasive division of political, social, cultural, and economic arrangements along language lines. The primary administrative regions of Belgium are based on linguistic patterns, and the structure of a wide array of institutions, ranging from the armed forces to the major political parties, reflects internal ethnoregional divisions. Even attitudes to issues such as education and economic development are often suggestive of regional sociolinguistic differences. These divisions are far more than simple reflections of language patterns in Belgium. They are the product of a complex history of social divisions that have assumed territorial significance. At the same time, they have served to alter fundamentally the geographical framework within which groups interact, accommodations are sought, and changes are initiated.

There is nothing natural or inevitable about the ways in which humans conceptually or formally divide territory for social or political purposes. Yet the causes and consequences of changing regional arrangements are of central importance to the evolution of group identity and intergroup communication. During the past century and a half, Belgium has evolved from a centralized state in which language patterns had no particular significance to a state comprising formally recognized language regions (since 1962) with a substantial degree of autonomy. The meaning of this transformation for the changing character of ethnolinguistic group identity and interaction is the central focus of this study.

The backdrop for the present ethnoregional divisions in Belgium is the essentially dichotomous linguistic geography of the country. Belgium straddles the border between the Romance and Germanic language families (figure 1). Since the early Middle Ages, the inhabitants of the southern part

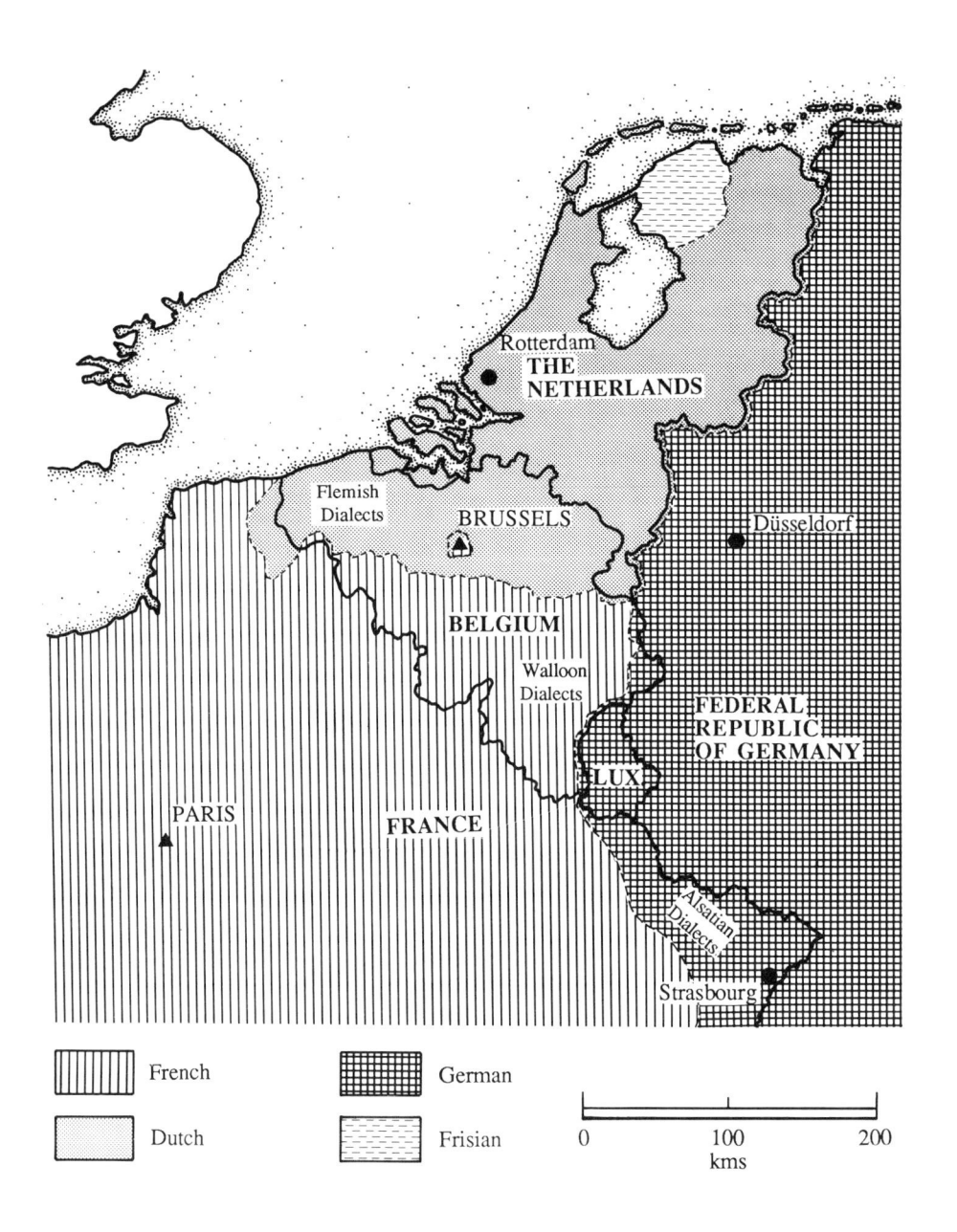

Fig. 1. Language regions of Northwest Continental Europe.

of what we now call Belgium have spoken Romance dialects closely akin to French, while the majority of those living in northern Belgium have used tongues related to Low German.[1] Through processes of language planning and standardization, French (français) has become the accepted language of educated writing and speech in the South, and Dutch (Nederlands) occupies a similar position in the North.[2] The inhabitants of northern Belgium outside of the Brussels metropolitan area have come to be referred to as Flemings (Vlamingen), and their region as Flanders (Vlaanderen). In the South, the inhabitants are now called Walloons (Wallons), and their region Wallonia (Wallonie). There is also a small German-speaking area in the eastern part of Belgium, largely as a result of the cession of the Eupen, Malmédy, and St. Vith districts to Belgium after World War I. Since this area contains less than 1 percent of Belgium's population, it has not figured prominently in Belgium's language problems and consequently will receive no extended consideration.

These generalizations about the language patterns of Belgium hide as much as they reveal. For a variety of historical reasons, at the time of Belgian independence a numerically small but politically and socially important Francophone minority lived in the northern part of Belgium, particularly in the larger urban centers. During the nineteenth and early twentieth centuries, this minority exerted a level of power and influence in the North greatly disproportionate to its size. In the capital city of Brussels the Francophones actually became a majority by the early decades of the twentieth century.[3] With the move toward regional unilingualism in recent decades,

---

[1] Historically the Germanic dialects of the North included Limburgisch, Flemish (which is sometimes divided into East and West variants), and Brabantic. These dialects are frequently lumped together under the term *Flemish*. The Romance dialects of the South, which are often referred to collectively as *Walloon*, included Walloon, Picard, Liègeois, and Lorrain. Hugo Baetens Beardsmore, "Linguistic Accommodation in Belgium," *Brussels Pre-Prints in Linguistics*, distributed by the linguistics circle of the Vrije Universiteit Brussel and the Université Libre de Bruxelles, 5 (March 1981), p. 7; Albert Verdoodt, "Introduction," *International Journal of the Sociology of Language* (Special issue on Belgium), 15 (1978), p. 5.

[2] The standard language of northern Belgium is sometimes referred to as Flemish. Given that there are only minor differences in written and spoken language between northern Belgium and the Netherlands, the preferred term for the common tongue of the Flemings is Nederlands, which is usually translated into English as Dutch. Moreover, the use of the term Flemish for the standard language of the northern part of Belgium is associated in the minds of some with Francophone attempts to demonstrate that the Flemings use a language of limited international significance. Hence, the term Dutch will be used to refer to the common language of the Flemings in this study. See generally, A. Van Loey, *La langue néerlandaise en pays flamand* (Brussels: Office de Publicité, 1945), p. 71; Aristide R. Zolberg, "Transformation of Linguistic Ideologies: The Belgian Case," in *Les états multilingues: problèmes et solutions*, ed. Jean-Guy Savard and Richard Vigneault (Quebec: Presses de l'Université Laval, 1975), p. 468, fn. 3.

the importance of the Francophone minority in most of Flanders has declined. In Brussels, by contrast, the French-speaking population has grown to the point that the percentage of Dutch speakers is estimated to be somewhere between 17.6 and 27 percent of the population.[4] Within Belgium's administrative structure, Brussels is officially a bilingual region, a reflection of the equal official status of the two main national languages in the capital rather than of the linguistic capabilities of most of its residents.

A remarkable feature of Belgium's linguistic geography is the relative numerical and territorial parity between Flemings and Walloons. Although differences in population and territorial control have become highly sensitive and politically charged domestic issues, there are few culturally or ethnically divided states that have as close a numerical balance between the major groups as does Belgium. Table 1 contains approximate percentages for the national language individuals responding to census questions indicated they used most frequently in 1846, 1910, and 1947. The figures should be interpreted with caution because of widely recognized problems with Belgian linguistic census data.[5] For one thing, there was only a limited range of available choices in response to language questions in the early censuses. Moreover, the changing nature of the language questions from one census to the next makes comparisons over time extremely difficult. In addition, infants were treated differently during the nineteenth century (when they were generally classified as speakers of the language of their parents or were not counted) and the twentieth century (when they were classified as speakers of none of the national languages). Finally, responses to language questions may not always have been entirely accurate due to both the great prestige of French at certain periods of Belgium's history and, in 1947, the political implications of language declarations. Nevertheless, table 1, provides some indication of the relative size of the language groups at three points in Belgium's history.

More recent censuses lack data on language use because language questions have been officially barred from the Belgian census since 1961.[6]

---

[3] *Wetenschappelijk Onderzoek van de Brusselse Taaltoestanden*, vol. 1 (Brussels: Nederlandse Commissie voor de Cultuur van de Brusselse Agglomeratie, 1974-1975), pp. 11-20.

[4] Albert Verdoodt, *Les problèmes des groupes linguistiques en Belgique*, Bibliothèque des Cahiers de l'Institut de Linguistique de Louvain, 10 (Louvain: Editions Peeters, 1977) p. 146. It is possible only to estimate the current number of Dutch speakers in the capital because the last Belgian census that included a question on language was conducted in 1947.

[5] See generally, Paul M. G. Levy, "Quelques problèmes de statistique linguistique à la lumière de l'expérience belge," *Revue de l'Institut de Sociologie*, 37, 2 (1964), pp. 251-273. The political implications of the 1947 census that are considered in chapter 6 probably resulted in a smaller percentage of persons claiming to use Dutch than was actually the case.

[6] Paul M. G. Levy, "La mort du recensement linguistique,"*Revue Nouvelle*, 36, 9 (1962), pp.145-154.

Table 1. Most Frequently Spoken Languages in Belgium, 1846-1947
(Percentages)

Year	French	Dutch	German
1846	42.1	57.0	0.8
1910	42.9	51.6	1.0
1947	41.9	52.9	0.9

NOTE: The totals for each year do not add up to 100 percent because the categories "other language" and "no response" have been omitted.

SOURCE: Calculated from: 1846: *Statistique générale de la Belgique. Exposé de la situation du Royaume (Période Décennale de 1841-1850)* (Brussels: Ministère de l'Intérieur, 1852), p. 18; 1910: *Annuaire statistique de la Belgique et du Congo Belge* (Brussels: Ministère de l'Intérieur, 1914), pp. 96-101; 1947: *Annuaire statistique de la Belgique et du Congo Belge* (Brussels: Ministère des Affaires Economiques, Institut National de Statistique, 1955), p. 99.

The only available statistics are population figures for the official language regions. Table 2 sets forth the percentages of the total population in each language region for 1970 and 1984. The relative linguistic homogeneity of language regions other than Brussels allows us to get some idea of the present balance between language groups. If Brussels is treated as being approximately 80 percent French speaking, calculations from the regional population data reveal a split between Dutch and French speakers of around 60 percent to 40 percent in 1984.[7] It is unlikely that the difference in percentages is any greater than this because there are almost certainly more Francophones in Flanders than Dutch speakers in Wallonia.[8]

Leaving aside for the moment issues of population trends, comparative social status, and political power, the Dutch speakers have always held a numerical edge over the Francophones in Belgium, but the edge has not

---

[7] The 80 percent figure for the French-speaking population of Brussels is based on a widely cited survey by Professors Kluft and Van der Vorst of the Institute of Sociology of the Free University of Brussels in 1969 which determined that 79.8 percent of the Brussels population used French in mental numerical calculations and that 78.9 percent declared a personal feeling of belonging to the French-speaking community. The figures for Dutch were 16.6 percent who use Dutch in mental calculations and 17.6 percent who felt a part of the Dutch-speaking community. Centre de Recherche de d'Information Socio-Politique,"L'évolution linguistique et politique du Brabant (I)," *Courrier Hebdomadaire du CRISP*, 466/467 (January 16, 1970), p. 11.

[8] Kenneth D. McRae, *Conflict and Compromise in Multilingual Societies: Belgium* (Waterloo, Ontario: Wilfrid Laurier University Press, 1986), pp. 280-281, estimates the present size of the Francophone population in Flanders at between 2 and 3 percent.

Table 2. *Population by Language Regions, 1970-1984* (Percentages)

Year	Dutch region	French region	Bilingual region	German region
1970	56.1	32.1	11.1	0.7
1984	57.5	31.9	10.0	0.7

SOURCE: Calculated from: 1970: *Recensement de la Population (31 décembre 1970)* vol. 1 (Brussels: Institut National de Statistique, 1984), p. 48; 1984: *Annuaire de Statistiques Régionales - 1984* (Brussels: Institut National de Statistique, 1984), p. 11.

been overwhelming. The reverse is true with respect to territorial considerations. Based on the current language regions, the French-speaking region (excluding the German-speaking district) occupies 52.4 percent of Belgium's total land area, whereas the Dutch-speaking region covers 44.3 percent of the country.[9] Thus, based on numerical and areal criteria neither the Walloons and their region nor the Flemings and their region is peripheral within the Belgian state.

Language patterns alone, of course, tell us very little about political and social divisions. In fact, linguistic heterogeneity in Belgium has not always translated into social divisions along language lines. Thus, many commentators have pointed to the lack of large-scale, well organized linguistically based group sentiment at the time of the Belgian Revolution in 1830.[10] Belgium began as a highly centralized state in which identities were either local, or in the case of a political and social elite, national. During the past 150 years, however, Belgium has seen the rise of a strong sense of group identity among many Flemings and Walloons, the development of significant tensions between the language communities, and the division of much of Belgium's political and social structure along language lines.

There is an enormous literature exploring the rise of language group consciousness in Belgium and the implications of that development.[11] The

---

[9] Calculated from *Annuaire de Statistiques Régionales* (Brussels: Institut National de Statistique, 1984), p. 8.

[10] See for example, Jean Stengers, "Belgian National Sentiments," in *Conflict and Coexistence in Belgium: The Dynamics of a Culturally Divided Society*, ed. Arend Lijphart (Berkeley: Institute of International Studies, 1981), pp. 46-60; Zolberg, "The Transformation of Linguistic Ideologies," pp. 450-451.

[11] The major bibliographies of Belgium's language problems fill several volumes. The most important of the recent bibliographies are: Verdoodt, *Les problèmes des groupes linguistiques en Belgique*; Albert Verdoodt, *Bibliographie sur le problème linguistique belge* (Quebec:

causes and consequences of what Aristide Zolberg termed "the making of Flemings and Walloons," have been the focus of intense scholarly investigation by, among others, historians, sociologists, and political scientists.[12] Although these inquiries have added considerably to our understanding of evolving social and political arrangements in Belgium, little attention has been devoted to a parallel process of vital significance, "the making of Flanders and Wallonia."[13] This process involved the creation of regional divisions in Belgium based on linguistic criteria of conceptual, and later functional, significance that fundamentally altered the context and nature of group identity and interaction. Although this process is sometimes recognized in the literature, it is rarely treated as more than a simple or inevitable result of language group differences.

The changing nature of regional conceptions of Belgium is symbolized by the history of the terms Wallonia and Flanders. At the time of the Belgian Revolution in 1830, there were no generally accepted toponyms for the language areas of Belgium. The only regional divisions shown on administrative maps from that period were the nine provinces, derived from the *départements* into which Belgium had been divided under French rule.[14] The term Wallonia had not yet been coined, and Flanders was used to refer either to the territory of the medieval county of Flanders or to the Belgian provinces of East and West Flanders.[15] At present, by contrast, the terms Flanders and Wallonia are used widely to denote the linguistically based administrative regions in Belgium of Dutch and French speech respectively. The terms have come to signify the territorial extent of language communities with common interests and a shared identity that, by virtue of institutional arrangements, enjoy a considerable degree of cultural, social, and economic autonomy.

International Center for Research on Bilingualism, 1983); and Lieve Suenaert and Pierre Verdoodt, "Langue et société en Belgique 1980-1985: bibliografie analytique et guide du chercheur/Taal en Maatschappij in Belgie 1980-1985: Analytische Bibliographie en Gids voor de Gebruiker," unpublished volume of the Centrum voor de Studie van de Pluriculturele Maatschappij/Centre pour l'Etude de la Société Pluriculturelle, 1986.

[12] Aristide R. Zolberg, "The Making of Flemings and Walloons: Belgium: 1830-1914," *Journal of Interdisciplinary History*, 5, 2 (Autumn 1974), pp. 179-235.

[13] See generally, Alexander B. Murphy, "Evolving Regionalism in Linguistically Divided Belgium," in *Nationalism, Self-Determination, and Political Geography*, ed. R. J. Johnston, David Knight, and Eleanor Kofman, (London: Croom Helm, 1987), pp. 135-150..

[14] Theo Luykx, *Atlas historique et culturel de la Belgique* (Brussels: Elsevier, 1959), plate 13.

[15] Albert Henry, *Esquisse d'une histoire des mots "Wallon" et "Wallonie"* (Brussels: La Renaissance du Livre, 1974); Maurits Gysseling, "Vlaanderen (Etymologie en Betekenisevolutie)," in *Encyclopedie van de Vlaamse Beweging*, ed. Jozef Deleu et al. (Tielt: Lannoo, 1975), pp. 1906-1912.

This study explores the evolution, nature, and implications of linguistic regionalism in an effort to demonstrate the significance of conceptual, functional, and formal (in the sense of politically instituted) compartmentalizations of territory for the development of group consciousness and intergroup relations. Belgium provides an excellent backdrop for such an inquiry. The use of linguistic criteria as a basis for subdividing the state has been both a response to growing language group consciousness and an agent in structuring intergroup relations. The language regions of Belgium are now at the heart of institutional accommodations to the language problem. Moreover, the types of language issues that have dominated the Belgian scene in recent decades are direct outgrowths of the establishment of formal linguistic regions. These include the existence of a Francophone minority in Flanders, the inclusion of the Fourons area in the Flemish region in 1962, and the containment of the spread of French in the area surrounding Brussels.[16]

The study begins with an examination in chapter 2 of the role of territory in the development of group consciousness and intergroup relations in plural societies. The concept of regionalism is introduced, and its significance as a dynamic force that reflects and shapes intergroup relations is developed. The conceptual framework is placed within the larger body of literature concerned with the spatial and territorial aspects of ethnic group relations. It is argued that important insights can be gained through an explicit focus on the reasons for, and implications of, changing regional/territorial conceptions and functions. Finally, methodological considerations are addressed with reference to the Belgian situation. The case is made for a central focus on the evolving language legislation of Belgium as a primary reflector and shaper of regional developments. The discourse employed by legislators and the leaders of the language movements is presented as a means of gaining insight into the reasons for territorial approaches to substate nationalist problems.

A study that is concerned with evolving conceptions of territory must necessarily be strongly historical in orientation. Chapter 3 introduces the historical dimension with an overview of the linguistic situation in the area we now call Belgium up until its independence in 1830. Since the focus of the study is on the role of changing regional conceptions and meanings within a politically organized area, this chapter provides the background necessary for the ensuing analysis. The argument is advanced that despite the importance of language in the formation of the Belgian state, the Ger-

---

[16] These three issues were selected as representative of Belgium's language problems in McRae, *Conflict and Compromise in Multilingual Societies: Belgium*, pp. 275-322.

manic and Romance language areas had no particular conceptual or political significance in 1830.

Belgium emerged as a highly centralized state with a French-speaking upper class.[17] French was the language of government, business, and higher education. Moreover, the dominant economic region of Belgium, the Sambre-Meuse Valley, was situated in the French-speaking part of the country. Over the course of the next seventy years, a Flemish movement gradually took form in reaction to the exclusion of the Dutch language at the upper levels of the Belgian government, economy, and society. Chapter 4 traces the emergence of this movement during the nineteenth century in relation to conceptions of territory and regionalism.

Historical evidence indicates that the primary thrust of the Flemish movement during this period was the attainment of the right for individual Flemings to use their language in the army, the courts, and the government.[18] These demands led to the enactment of the first important language laws during the last quarter of the nineteenth century.[19] By the end of the nineteenth century, a Walloon movement was also beginning to take form in response to growing Flemish successes. Given the dominant position of French in Belgium at the time, however, there was little impetus for any significant developments on the Walloon front.

The central proposition of chapter 4 is that the present regions of Flanders and Wallonia had little conceptual or functional integrity in the nineteenth century. The Dutch language area had acquired some degree of conceptual significance among certain segments of Flemish society by 1900, but Flemish regional consciousness was still quite limited throughout most of northern Belgium, and Walloon regional identity was virtually nonexistent in the southern part of the country. This raises questions about the all too frequent practice of viewing Flanders and Wallonia as analytical units for analyzing developments during the nineteenth century. This issue is explored in some detail in an effort to demonstrate the pitfalls of projecting present regional structures into the past in our efforts to explain the rise of intergroup tensions.

An important change occurred in the Flemish movement, and even to some degree in its nascent Walloon counterpart, during the early decades

---

[17] See generally, Janet Polasky, "Liberalism and Biculturalism," in *Conflict and Coexistence in Belgium*, ed. Lijphart, pp. 34-45.

[18] Shepard B. Clough, *A History of the Flemish Movement in Belgium: A Study in Nationalism* (New York: Richard R. Smith, 1930), pp. 130-171.

[19] For a survey of the language laws, see Pierre Maroy, "L'évolution de la législation linguistique belge," *Revue du Droit Public et de la Science Politique*, 82, 3 (May -June 1966), pp. 457-460.

of the twentieth century. There was a shift in focus from an emphasis on the rights of individual language speakers to the collective rights of the inhabitants of the language regions.[20] Chapter 5 analyzes this transformation, which introduced a territorial approach to language group issues and laid the foundation for the later divisions of the state. The leaders of the Flemish movement called for cultural autonomy for Flanders and a unilingual policy in the region both as a means of gaining leverage in a political context in which they were at a disadvantage and as a strategy for promoting Flemish identity throughout northern Belgium. In reaction to the gains of the Flemings, certain Walloons began calling for similar measures for Wallonia. Works began appearing on Flanders and Wallonia, Fleming and Walloon congresses addressed issues of regional autonomy, and Flemish and Walloon activists began promoting the traditional symbols of nationalism: flags, songs, and slogans. Flanders and Wallonia were presented as natural territorial units, and the creation of regions as ideologies was underway.

Nevertheless, until World War I only the more radical fringes of the linguistic nationalist movements espoused and supported the idea of language regions as cultural and political units to which power should be devolved from the centralized Belgian state. Developments during the war, including the introduction by the occupying Germans of an administrative partition of Belgium along language lines in an effort to capitalize on internal divisions, served to promote the conception of Flanders as a distinct territorial entity despite the postwar rejection of German influence. The growth of a linguistically based regional conception of Belgium during the early twentieth century culminated in the enactment of a series of important language laws in the 1930s giving formal expression to these territorial conceptions. Chapter 5 traces these developments and explores the role of the 1930s laws in altering regional conceptions and patterns in Belgium.

Although the territorial legislation of the 1930s promoted thinking about Belgium in terms of language regions, the laws brought about only limited changes. There were no enforcement mechanisms or sanctions, and the laws met with widespread resistance among French speakers. The resulting disillusionment of many Flemings, together with mounting concern among Walloons during the 1950s over the growing economic and numerical dominance of the Flemings within Belgium, set the stage for the sweeping restructuring of the Belgian state along language lines that has taken place during the past twenty-five years. Chapter 6 examines the legislative and constitutional provisions that have been enacted in recent

---

[20] See generally, Val R. Lorwin, "Linguistic Pluralism and Political Tension in Modern Belgium," *Canadian Journal of History*, 5, 1 (March 1970), p. 13.

decades and outlines the most important changes that have been introduced in political, social, cultural, and economic arrangements. This chapter provides the background for an inquiry in the following chapter into the implications of this profound institutionalization of linguistic differences. In addition, it explores the reasons for those changes. Particular attention is paid to the role of prior regional developments in focusing language group concerns on territorial issues and the importance of feelings of disadvantage and oppression by both Flemings and Walloons in encouraging collective regional approaches to issues of government and society.

Chapter 7 takes up the central issue of the implications of regional developments in Belgium. Attention is devoted to the ways in which the creation of formal language regions has affected the political structure of Belgium, the flow and presentation of information, the structure of public and private institutions, the specific issues of contention between language groups, the ways in which intraregional differences are perceived and understood, the degree of group cohesion and intergroup contact, the relative importance of language issues in Belgian politics, and a number of other issues related to demographics, economics, and urban structure. Any one of these issues could, of course, be the subject of a full length study by itself. Moreover, the comparative recency of the most sweeping formal regional developments makes an extensive analysis of consequences problematic at best. Hence, the goal of this chapter is to identify the range of relevant factors in an effort to enhance our understanding of the potential ramifications of ethnoregionalism in a plural society.

The concluding chapter draws together the issues that have emerged in the course of the inquiry into the Belgian case in an effort to demonstrate the analytical significance of an explicit focus on regional/territorial developments. Emphasis is placed on: (1) the importance of viewing regional developments as dynamic, powerful forces that reflect and shape social arrangements; (2) the perils of assuming that present regional divisions can be projected into the past in efforts to construct causal explanations for the rise of intergroup tensions; (3) the powerful role of formal, institutionalized regionalism in shaping human spatial organization and perception; and (4) the nature of the insights that can be gained from a focus on the effects of regional developments on the evolution of group consciousness and intergroup relations. Building on the last point, the conclusion points to some of the advantages and drawbacks of the formal regional arrangements that have been instituted in Belgium.

The institutionalization of ethnolinguistic differences in formal regional arrangements in modern Belgium is having an important impact on the human geography of that country. Although some geographers and others have sought to understand certain aspects of the situation, the role of territorial developments in reflecting and shaping the language problems of Belgium has yet to be articulated. In other words, there is a need for a broad but fundamental consideration of how and why politically and functionally important regional divisions of an ethnolinguistic nature came about, and of the present and future significance of those regional developments. This study attempts to take a step in that direction.

Chapter 2

ETHNICITY, REGIONALISM, AND THE SUBSTATE
NATIONALIST QUESTION

The political pattern of the world bears remarkably little relationship to more complex patterns of cultural or ethnic identity. The term *nation-state*, in its classical usage, is based on the idealized notion of territorial correspondence between sovereign political entities and groups sharing a common sense of political and cultural identity.[1] In reality, however, virtually no modern state fits this conception,[2] and internal strife between groups with differing cultural and political orientations has been one of the most frequent sources of conflict since World War II.[3] In response to the obvious significance of this phenomenon, interest in problems of culture, ethnicity, and nationalism has surged in the past two decades, and there is now an extensive body of literature on the subject.[4] Much of the published material is descriptive and case specific, but substantial efforts have been made to explore broader issues and to provide a stronger theoretical and analytical basis for studies of culture and nationality.

---

[1] The classical usage of the term is based upon an understanding of "nation" as a cultural unit and "state" as a formal political unit. See generally, Walker Connor, "A Nation is a Nation, is a State, is an Ethnic Group, is a . . . ," *Ethnic and Racial Studies*, 1, 4 (October 1978), p. 388; Norman J. G. Pounds, *Political Geography* (2nd ed.; New York: McGraw-Hill, 1972), pp. 3-17. The definition and significance of these terms will be explored in more detail below.

[2] See generally, Marvin W. Mikesell, "The Myth of the Nation State," *Journal of Geography*, 82, 6 (November-December 1983), pp. 257-260.

[3] Richard D. Lambert, Preface to "Ethnic Conflict in the World Today," *The Annals of the American Academy of Political and Social Science*, 433 (September 1977), p. vii.

[4] The relative recency of this interest is suggested by Walker Connor's lamentation in a 1972 article of the lack of attention devoted to ethnic groups and nationalism in the political science literature. Walker Connor, "Nation-Building or Nation-Destroying?" *World Politics*, 24, 3 (April 1972), p. 319.

The wide array of relevant concepts, the breadth and ambiguity of many of the issues, and the lack of consensus on definitions and terminology pose a number of problems for the scholar working in this area. The situation is further complicated by the variety of disciplinary approaches and the frequent lack of cross-fertilization of ideas.[5] In seeking to orient the reader, this chapter takes up definitional issues first, discussing the nature of cultural traits, ethnic groups, and nationalism, and their relation to one another. This is followed by a brief overview and critical assessment of the types of approaches that have been adopted to explain the evolution of substate nationalist movements. The concepts of regionalism and territoriality are then introduced and their analytical significance explored. Attention is devoted to the relation between the approach adopted in this study and the geographical literature on culture and nationality. Finally, the Belgian case is presented as a vehicle for investigating the significance of evolving regional/territorial conceptions in the rise of substate nationalism. Methodological justification is given for an emphasis on the role of law in creating and reflecting ethnoregional divisions and on the discourse of elites in exposing the reasons for, and conceptual significance of, territorial developments.

Culture, Ethnicity and Nationalism

The literature on ethnicity and nationalism is characterized by considerable confusion of word usage.[6] As a result, much of the careful scholarship on the subject devotes substantial attention to questions of terminology. A term frequently employed in contemporary scholarship on culture and nationality is *ethnicity*. Ethnicity is a comparatively new term that has been used in a variety of ways.[7] Although it was once understood

---

[5] Walker Connor points to the lack of overlap between the contributions to two of the leading journals concerned with aspects of culture and nationality, the *Canadian Review of Studies in Nationalism* and *Ethnic and Racial Studies*, despite the fact that the two journals publish articles on similar issues. "A Nation is a Nation," p. 387.

[6] See, for example, Fred W. Riggs, "What is Ethnic? What is National? Let's Turn the Tables," *Canadian Review of Studies in Nationalism*, 13, 1 (Spring 1986), pp. 111-123. In an effort to deal with the confusion, a glossary of terms has been compiled based on scholarly usage of terms in "ethnicity research," with plans to supplement and update the work at regular intervals. Fred W. Riggs, ed., *Ethnicity, INTERCOCTA Glossary, Concepts and Terms used in Ethnicity Research*, International Social Science Council Committee on Conceptual and Terminological Analysis (Pilot ed.; Honolulu: Department of Political Science, University of Hawaii, n.d.).

[7] The word ethnic is derived from the Greek word *ethnos* meaning people, nation, or country. Benjamin Akzin, *State and Nation* (London: Hutchinson University Library, 1964), p. 36. Given the breadth of the Greek term, it is not surprising that there is considerable confusion

as a synonym for race, it has now taken on a strong cultural dimension.[8] Frederik Barth notes that many anthropologists accept a definition of an ethnic group as a set of people that "(1) is largely biologically self-perpetuating; (2) shares fundamental cultural values, realized in overt unity in cultural forms; (3) makes up a field of communication and interaction; and (4) has a membership which identifies itself, and is identifiable by others, as constituting a category distinguishable from other categories of the same order."[9]

As Barth points out, the problem with definitions of ethnicity of this sort is that they put objective cultural criteria to the fore, thereby emphasizing features that may be more the result than the cause of the development of units of social organization.[10] Moreover, a preoccupation with objective cultural traits tells us little about the significance of these traits in different social contexts. An alternative is to place primary emphasis on self-awareness, a subjective approach to ethnicity that has gained wide acceptance.[11] Focusing on the self-ascriptive aspect of ethnicity draws attention to the idea that ethnic groups must be understood in relation to other peoples and institutions around which feelings of distinctiveness and identity have developed.[12] The concept thus involves a "we-they syndrome" that requires the presence of "relevant others."[13]

---

surrounding the concept of ethnicity, a problem compounded by the failure of some scholars to define their use of the term carefully. Wsevolod Isajiw, "Definitions of Ethnicity," *Ethnicity*, 1 (July 1974), pp. 11-24.

[8] Louis L. Snyder argues that giving ethnicity a cultural dimension is an unjustified and potentially dangerous departure from the traditional understanding of ethnicity as "the physical and mental traits in races." "Nationalism and the Flawed Concept of Ethnicity," *Canadian Review of Studies in Nationalism*, 10, 2 (Fall 1983), pp. 253-265. Snyder himself admits, however, that the concept has been extended well beyond the racial aspect, and many others assert that this is entirely justified in view of the etymology of the term and the need for a word to describe particular types of cultural groupings. See, for example, Riggs, "What is Ethnic?" pp. 111-123.

[9] Frederik Barth, ed., *Ethnic Groups and Boundaries: The Social Organization of Culture Difference* (Boston: Little, Brown & Co., 1969), pp. 10-11.

[10] *Ibid.*, pp. 11-12.

[11] For example, Harrold Isaacs, *Idols of the Tribe. Group Identity and Political Change* (New York: Harper & Row, 1975); Immanuel Wallerstein, "Ethnicity and National Integration," *Cahiers d'Etudes Africaines*, 1, 3 (July 1960), pp. 129-139.

[12] Bernard E. Segal, "Ethnicity: Where the Present is the Past," in *Ethnic Autonomy: Comparative Dynamics, the Americas, Europe, and the Developing World*, ed. Raymond L. Hall (New York: Pergamon Press, 1979), p. 10.

[13] Crawford Young, *The Politics of Cultural Pluralism* (Madison, Wisconsin: University of Wisconsin Press, 1979), p. 42.

Despite the obvious significance of this relational view of ethnicity for studies of group interaction, ethnic groups should not be thought of wholly in terms of subjective criteria.[14] To do so is to risk blurring all distinctions between ethnic groups and other social units. After all, the modern use of the term ethnicity came into being out of a need to describe the special relations that exist in certain contexts between peoples of similar race, culture, or common ancestry. Although the subjective element of ethnicity should be stressed, it should be tied to cultural criteria. Hence, in this study ethnicity refers to the ascription of group distinctiveness based on the perception of common culture (as evidenced by such factors as language or religion), racial similarity, and/or shared ancient ancestral origin that serve to distinguish two or more groups living in close proximity to one another. An ethnic group, then, is a unit of social organization defined on the basis of its ethnicity.

Before leaving this definition, a few clarifications are necessary. First, the definition does not specify that perceptions of distinctiveness must be self-ascribed, since ethnic identity can be ascribed by others.[15] In the Belgian context, for example, there are many individuals within Flanders who do not attach any particular significance to a wider Fleming ethnic identity, yet they are effectively accorded that identity for certain purposes under the laws of Belgium. This is suggestive of two important related points: that ethnic groups are not homogeneous, and that members of ethnic groups possess multiple forms of identity that need not be mutually exclusive and that, for some, do not necessarily include identification with the relevant ethnic group.[16] Moreover, the emphasis in the definition on ascription and perception points to the fluid nature of ethnicity as a phenomenon that changes depending upon context and point of view. Hence, ethnicity should be understood as a process rather than an objective and immutable state. Finally, the term ethnic is frequently combined with certain adjectives to form compounds such as ethnolinguistic, ethnoregional, and ethnonationalist. This study employs these forms as a means of defining more specifically the nature and character of ethnicity.[17]

---

[14] Pierre Van den Berghe has criticized Barth for overemphasizing social boundaries at the expense of cultural criteria. "Ethnic Pluralism in Industrial Societies: A Special Case?" *Ethnicity*, 3 (1976), p. 242.

[15] Cf. Cynthia H. Enloe, *Ethnic Conflict and Political Development* (Boston: Little, Brown & Co., 1973), pp. 17-18.

[16] Robert G. Wirsing, ed., *Protection of Ethnic Minorities: Comparative Perspectives* (New York: Pergamon Press, 1981), p. 7.

[17] For example, the term ethnolinguistic is used to convey the idea of ethnic identity as defined primarily in linguistic terms.

Consideration must also be given to the concept of nationalism. The term *nation* has so many different connotations that it is easily confused either with state or with ethnic group.[18] The distinction between a nation and a state is based on the generally accepted notion that a nation is a particular social grouping, whereas a state is a sovereign politically organized area.[19] Distinguishing a nation from an ethnic group is much more difficult because both concepts are based on units of social organization.[20] Anthony Smith contends that the nation is the modern outgrowth of the ethnic community.[21] The distinguishing feature of nationality is its political dimension.

Defining a nation in political terms is a logical extension of the generally accepted understanding of modern nationalism as a sentiment that arose in the eighteenth century in connection with rationalism and the political upheavals surrounding the French Revolution.[22] The underlying ideology holds that a group of people has the right to govern itself. This notion is embodied in most definitions of nation or nationalism that make reference to a group's self-conscious quest for, or achievement of, a degree of control over its own affairs, by which is usually meant the ability of group members or their representatives to make decisions about local political, cultural, and social arrangements affecting the group.[23]

Before attempting an exact definition here, a word should be said about the relation between ethnicity and nationalism. Criticism has been directed at such scholars as Hans Kohn and Karl Deutsch for failing adequately to explore the role of ethnicity in modern nationalism.[24] Walker

---

[18] Isaacs, *Idols of the Tribe*, p. 27; Anthony D. Smith, *The Ethnic Revival in the Modern World* (Cambridge: Cambridge University Press, 1981), p. 85.

[19] For example, Pounds, *Political Geography*, pp. 1-11.

[20] In fact, the derivation of the term *nation* invites confusion with ethnicity because nation, which comes from the Latin word *nasci* meaning "to be born" implies common ancestry. Leonard Tivey, "States, Nations and Economies," in *The Nation-State: The Formation of Modern Politics*, ed. Leonard Tivey (Oxford: Robertson, 1981), pp. 4-5.

[21] Anthony D. Smith, "Ethnic Identity and World Order," *Millennium: Journal of International Studies*, 12 (1982), pp. 149-161.

[22] See generally, Hans Kohn, *The Idea of Nationalism: A Study in its Origins and Background* (New York: Macmillan, 1944).

[23] For example, Hans Gerth and C. W. Mills, *Character and Social Structure* (New York: Harcourt, Brace & World, 1953), p. 197. They define a nation as "a body of people which by cultural traditions and common historical memories is capable of organizing a state, or at least which raises the claim for such an autonomous organization with some chance of success."

Connor persuasively argues that ethnic groups are the building blocks of nations.[25] He points to the idea that nations possess the same feelings of group attachment that characterize ethnic groups, and contends that nations are simply ethnic groups that have become politically aware.

The ethnic dimension of modern nationalism is captured in the comprehensive definition of a nation offered by a prominent commentator on the subject, Konstantin Symmons-Symonolewicz. He defines a nation as "a territorially based community of human beings sharing a distinct variant of modern culture, bound together by a strong sentiment of unity and solidarity, marked by a clear historically-rooted consciousness of national identity, and possessing, or striving to possess, a genuine political self-government."[26] This definition embodies the significant aspects of nationality discussed above, and can therefore serve well for the present study with one modification.

The importance of nationalism for state formation sometimes draws attention away from the political activities of ethnic groups within states that are seeking a degree of autonomy that may fall short of complete independence. Yet this process also involves the attempt by ethnic groups to gain control over their own affairs, albeit in the more limited sense of control over matters such as education, cultural affairs, and aspects of local government.[27] Given the possible interpretation of the phrase "genuine self-government" in the Symmons-Symonolewicz definition as a reference only to sovereign state status, the last clause could be modified to encompass groups "possessing, or striving to possess, genuine self-government or a substantial degree of control over their own political and cultural affairs at a local or regional level." Nationalism, then, refers to the ideological expression of the goals of a nation, and to the processes by which those goals are pursued.

---

[24] Ken Wolf, "Ethnic Nationalism: An Analysis and a Defense," *Canadian Review of Studies in Nationalism*, 13, 1 (Spring 1986), pp. 99-109.

[25] Connor, "A Nation is a Nation," pp. 377-400; "Ethnic Nationalism as a Political Force," *World Affairs*, 133 (1970), pp. 91-97.

[26] Konstantin Symmons-Symonolewicz, "The Concept of Nationhood: Toward a Theoretical Clarification," *Canadian Review of Studies in Nationalism*, 12, 2 (Fall 1985), p. 221.

[27] Alan Butt Philip, for example, defines nationalism as "the active solidarity of a group of people who share a common culture or history and a sense of nationhood and who seek to give this common experience a political reality by means of self-government or some other kind of political recognition, if not autonomy." "European Nationalism in the Nineteenth and Twentieth Centuries," in *The Roots of Nationalism: Studies in Northern Europe*, ed. Rosalind Mitchinson (Atlantic Highlands, New Jersey: Humanities Press, 1984), pp. 1-2.

Nationalism, of course, takes on many forms, and attempts have been made to identify its various types.[28] Although it is beyond the scope of the present inquiry to consider these types in detail, it is important to clarify the orientation of this study. The concern here is with the aspirations of ethnic groups living within states to obtain a degree of control over local or regional political and cultural matters. The term "substate nationalism" describes this situation and serves to distinguish it from nationalisms involving state expansion (for example, Germany during World War II), unification of multiple politically organized areas into one state (for example, the formation of modern Germany and Italy), liberation from colonial domination (for example, most of the Latin American and African states), or other larger scale varieties of nationalism.[29] Before exploring the particular orientation of this study, it is necessary to say a few words about other approaches to the rise of substate nationalism.

Perspectives on the Rise of Substate Nationalism

The study of substate nationalism raises questions concerning the causes for both the initial development of ethnic groupings and the attainment by those groups of political objectives. It is impossible in a limited space to discuss the full range of ideas that have been developed to explain substate nationalism in such diverse contexts as former colonial states, industrialized Western democracies, and Socialist states. Hence, the emphasis here will be on studies of the rise of substate nationalism in Western Europe.

Most West European states, as so-called stable, mature, developed, and modernized political entities, have a relatively high standard of living, a well developed infrastructure promoting social interaction, a democratic tradition of government, and an array of political institutions designed to promote internal stability. The West European democracies did not emerge from the kind of recent colonial experience that had profound impacts on ethnic group relations throughout so much of the rest of the world.[30] In

---

[28] Two of the more recent typologies are Andrew W. Orridge, "Varieties of Nationalism," in *The Nation-State: The Formation of Modern Politics*, ed. Tivey, pp. 39-58; and Young, *The Politics of Cultural Pluralism*, pp. 92-97.

[29] The term minority nationalism has been used by some to convey the same meaning as substate nationalism does here. For example, Colin H. Williams, "Perspectives on Minority Nationalism in Europe," in *Nationalism in the Modern World*, ed. John M. Walters (London: Croom Helm, 1986). The term substate nationalism is preferable for this study because at particular stages in Belgium's history, neither the Flemings or Walloons could properly be called minorities, even though they may have felt disadvantaged.

[30] Much of the important early work on ethnic relations within states came out of studies of colonial situations. Of particular note are the works of Furnivall and Boeke on Southeast

addition, the idea of the nation-state, a distinctly Western notion, permeated the development of most modern West European states to a degree experienced by few other areas in the world. Moreover, the democratic political traditions of Western Europe provide a framework for political action that is distinctly different from traditions prevailing in many other states. Finally, the significant intensification of substate nationalist feelings in many West European states during the 1960s and 1970s drew considerable attention to problems ranging from the highly visible and violent separatist movements in Northern Ireland and the Basque country to the more moderate, but equally persistent, ethnonationalist struggles of the Bretons in France, the Flemings and Walloons in Belgium, and the Welsh in the United Kingdom.[31] This has led to the development of a significant body of literature in recent years on substate nationalism in this part of the world.

The upsurge in visibility of the substate nationalist movements in Western Europe came as a surprise to many who, following the relatively quiescent period during the decade and a half after World War II, assumed that the Western democracies had either solved, or were rapidly in the process of solving, ethnic conflict through modernization and assimilation. This line of thinking was profoundly influenced by the writings of Karl Deutsch. Deutsch's basic argument was that modernization serves to mobilize and put into contact large segments of the population, thus promoting homogeneity.[32] Subsequent developments have cast significant doubt on this thesis. Arend Lijphart argues that the theory is inconsistent and its interpretation oversimplistic.[33] He asserts that Deutsch and his followers fail to take into account the range of factors necessary for assimilation and the differential effects of modernization at successive stages of development.[34]

Asia and of Smith on the Caribbean. J. S. Furnivall, *Colonial Policy and Practice* (Cambridge: Cambridge University Press, 1948); J. H. Boeke, *Economics and Economic Policy of Dual Societies* (New York: Institute of Pacific Relations, 1953); M. G. Smith, "Social and Cultural Pluralism," *Annals of the New York Academy of Sciences*, 83, 5 (1960), pp. 786-795.

[31] Surveys of these problems can be found in Meic Stephens, *Linguistic Minorities in Western Europe* (Llandysul, Wales: Gomer Press, 1976); and Charles R. Foster, ed., *Nations Without a State: Ethnic Minorities in Western Europe* (New York: Praeger Press, 1980).

[32] Karl W. Deutsch, *Nationalism and Social Communication: An Inquiry into the Foundations of Nationality* (Cambridge: Technology Press of M.I.T., 1953). This view was not shared by some previous commentators on the implications of modernization. For example, Carlton Hayes, *The Historical Evolution of Modern Nationalism* (New York: Macmillan, 1931), pp. 234-237.

[33] Arend Lijphart, "Political Theories and the Explanation of Ethnic Conflict in the Western World: Falsified Predictions and Plausible Postdictions," in *Ethnic Conflict in the Western World*, ed. Milton J. Esman (Ithaca: Cornell University Press, 1977), p. 48.

With the rise of substate nationalism in Western Europe during the past two decades, the significance of ethnicity as a force in the modern state can hardly be denied.[35] Yet, as Mark Kauppi points out, the three major theoretical perspectives on state-society relations—classical liberalism, classical Marxism, and organic-statism—fail to recognize ethnic groups as significant social units within modern political systems.[36] Kauppi argues that liberalism's stress on individualism, social atomism, and competition ignores group cooperation and communal identity.[37] Classical Marxism sees ethnic groups as anachronisms from the pre-capitalist period of history that become less significant with the rise of class interests, and eventually lose all meaning in the transition to communism.[38] In fact, Marx and Engels had little patience for ethnic groups, referring at one point to such groups as the Basques, Bretons, and Gaels as "national refuse."[39] Finally, the focus of organic-statism on the importance and naturalness of the political community as a whole largely ignores ethnic segmentation.[40] The failure of any of these theories to take ethnic group identity seriously has arguably impeded our understanding of the ethnic factor in national integration.

The inchoate state of conceptual development concerning the rise of substate nationalism in Western Europe renders the task of assessing theoretical contributions rather difficult. By focusing on research objectives rather than the confusing array of epistemological perspectives and systematic orientations, however, it is possible to identify three general

---

[34] Lijphart finds other reasons for the mistaken expectation that ethnicity would decline in significance in the West, including liberal wishful thinking, the favorable experience of the West in comparison with colonial states, the popularity of the "end of ideology" thesis, the emphasis on supranational integration in Europe, and Marxist overemphasis on class and the influence of materialism on human affairs. *Ibid.*, pp. 49-55.

[35] This point is explored in detail in the introduction to Nathan Glazer and Daniel P. Moynihan, eds., *Ethnicity: Theory and Experience* (Cambridge, Massachusetts: Harvard University Press, 1975).

[36] Mark V. Kauppi, "The Resurgence of Ethno-Nationalism and Perspectives on State-Society Relations," *Canadian Review of Studies in Nationalism*, 11, 1 (Spring 1984), pp. 119-132.

[37] *Ibid.*, pp. 122-123. See also Vernon Van Dyke, "The Individual, the State and Ethnic Communities in Political Theory," *World Politics*, 29, 3 (April 1977), pp. 343-369.

[38] A major review of Marxism and ethnicity is Walker Connor, *The National Question in Marxist-Leninist Theory and Strategy* (Princeton: Princeton University Press, 1984). See also, Frank Parkin, *Marxism and Class Theory: A Bourgeois Critique* (London: Travistock, 1979), pp. 4-5.

[39] Frederick Engels, "The Magyar Question," quoted in Karl Marx, *The Revolutions of 1848*, Political Writings, vol. 1, ed. and translated by David Fernbach (London: Allen Lane in association with New Left Review, 1979), pp. 221-222.

[40] Kauppi, "The Resurgence of Ethno-Nationalism," pp. 126-127.

approaches in the theoretical literature on ethnicity and nationalism in Western Europe: (1) attempts to construct overall theoretical frameworks (for example, internal colonialism, center-periphery relations); (2) attempts to identify and assess the role of widespread historical transformations (for example, modernization, centralization); and (3) attempts to identify and assess the role of particular internal social, institutional, or behavioral processes and structures (for example, legal arrangements, elite behavior). There is, of course, considerable overlap in the issues addressed by the contributors to this literature. As a result, focusing on the nature of the theoretical inquiry rather than the issues themselves may be the only feasible approach to orienting new work within the conceptual morass.[41]

The perspective of the present study belongs within the third category of approaches identified above, but setting it within the context of all three is desirable. Among the attempts to develop overall explanatory frameworks or models of substate nationalism, the essentially materialist conceptions of Hechter on "internal colonialism" and Nairn on "uneven development" have been particularly widely discussed. Both are derived from notions of imperialism and dependency. Hechter, using the United Kingdom as an example, argues that states become integrated through a process of expanding domination from the core, thereby relegating the periphery to a permanently disadvantaged or exploited status from an economic and cultural standpoint.[42] Ethnonationalist movements form as a reaction to this "internal colonialism" during periods in which the power of the core is in decline. Nairn sees ethnic separatism as a consequence of uneven development resulting from the functioning of the world capitalist economy.[43] Arguing from a Marxist perspective, Nairn contends that the inevitable exploitation and domination of some regions by others that accompanies capitalist expansion generates social fragmentation into competing ethnic units.

Although these perspectives serve to focus attention on important economic aspects of substate nationalism, they have been widely criticized on empirical and theoretical grounds.[44] Empirical problems lie in the

---

[41] Attempts to assess the state of conceptual development in this field frequently end up being more confusing than helpful, in part because of the virtual impossibility of organizing contributions in terms of thematic or epistemological approach.

[42] Michael Hechter, *Internal Colonialism: The Celtic Fringe in British National Development, 1536-1966* (Berkeley: University of California Press, 1975).

[43] Tom Nairn, *The Break-up of Britain* (London: New Left Books, 1977).

[44] For a summary of criticisms to these conceptions, see Andrew W. Orridge and Colin H. Williams, "Autonomist Nationalism: A Theoretical Framework for Spatial Variations in its Genesis and Development," *Political Geography Quarterly*, 1, 1 (January 1982), p. 35;

failure of the theories to tell us much about the timing, location, or intensity of ethnonationalism. Thus, many of the areas in question are not notably underdeveloped (for example, the Basque country); nationalist activity has frequently not coincided with economic downturns (for example, recent Flemish nationalism); and more developed areas are not necessarily more tranquil than less developed areas (for example, Catalonia as opposed to Galicia). In addition, the theories posit, but fail to demonstrate, that capitalistic development follows ethnic lines. Furthermore, these approaches provide little insight into why people come to think of themselves in nationalist terms. In fact, differences in economic fortune are clearly insufficient by themselves to account for substate nationalism, as indicated by Milton Esman's observation that "while economic conditions in Northern England have been as bleak as those in neighboring Scotland, they have not resulted in the organization or expression of politically significant grievances."[45] Finally, both John Agnew and Phillip Rawkins have criticized Hechter and Nairn for the undue emphasis they place on economic structures at the expense of cultural and political factors.[46] As Rawkins points out, the overwhelming weight attached to structural factors eliminates the realm of experiences from ethnic relations by diverting attention away from how people comprehend structural differences and accommodate to alternative cultural identities.[47]

The internal colonialism and uneven development theories are outgrowths of a core-periphery view of socio-political change.[48] In view of the broad range of ideas that are subsumed within the core-periphery perspective, it is difficult to form generalizations about them. As Agnew points out, however, theories of this sort, when applied to questions of substate nationalism, logically predict a uniform commitment to nationalism in the periphery whereas, in fact, ethnic groups tend to be

---

Anthony D. Smith, "Nationalism, Ethnic Separatism and the Intelligentsia," in *National Separatism*, ed. Colin H. Williams (Vancouver: University of British Columbia Press, 1982), pp. 20-24.

[45] Milton J. Esman, "Perspectives on Ethnic Conflict in Industrial Societies," *Ethnic Conflict in the Western World*, ed. Esman, p. 377.

[46] John A. Agnew, "Structural and Dialectical Theories of Political Regionalism," in *Political Studies from Spatial Perspectives*, ed. Alan D. Burnett and Peter J. Taylor (New York: John Wiley & Sons, 1981), pp. 275-289; Phillip Rawkins, "Nationalist Movements within the Advanced Nationalist State: The Significance of Culture," *Canadian Review of Studies in Nationalism*, 10, 2 (Fall 1983), pp. 221-233.

[47] Rawkins, "National Movements within the Advanced Nationalist State," p. 230.

[48] For a general discussion of center-periphery ideas, but with examples ranging well beyond the problems of ethnonationalism, see Jean Gottmann, ed., *Centre and Periphery: Spatial Variations in Politics* (London and Beverly Hills: Sage, 1980).

highly heterogeneous with regard to political ideology and perceptions of identity.[49] Finally, center-periphery conceptions provide little helpful insight into the problems of such polycentric states as Belgium, Switzerland, or Yugoslavia.

The emphasis in all of these explanatory frameworks on structural factors prompted Agnew to argue for a dialectical theory based on the relation between human behavior and objective structure. A dialectical approach seeks to identify the processes that link human actions and social structures in an attempt to understand "how people define their interests and settle on the strategies which produce their political behavior."[50] Agnew persuasively argues the advantages of viewing substate nationalism as the result of human action under conditions of structural constraint, but the problem remains as to whether one can construct an overall explanatory framework for a process that occurs in widely different geographical, social, and political contexts. Moreover, even a short listing of factors involved in the rise of substate nationalism would have to include the conditions under which peoples sharing common cultural or racial traits and/or ancient ancestral origin acquire self-awareness as a social unit, the relative size and position of ethnic groups within the state, the extent and type of contact among groups, the nature and depth of cultural and ideological differences, the degree of internal group cohesiveness, the importance of disparities in economic and political power, and the role of institutions (particularly the state) in structuring group relations.[51] In light of such a wide range of relevant issues, tying an explanation to any one model or conception of social process risks being reductionist.

A different approach represented in the literature involves efforts to identify and examine the role of widespread historical transformations in precipitating ethnic separatism. Among the principal developments that have been considered are centralization, modernization, interdependency, and decolonization. Lijphart asserts that, contrary to Deutsch's idea that modernization fosters assimilation, the rapid increase in social communication associated with modernization frequently causes a lag or even a decrease in assimilation.[52] Moreover, he contends that the growth of

---

[49] Agnew, "Structural and Dialectical Theories of Political Regionalism," pp. 278-279.

[50] *Ibid.*, p. 283.

[51] A number of these variables are suggested in Lawrence Osinski, "Multi-Ethnicism and the Nation-State in Southeast Asia," (M. A. thesis, Department of Geography, The University of Chicago, 1971), pp. 29-35. See also, Milton M. Gordon, "Toward a General Theory of Racial and Ethnic Group Relations," in *Ethnicity: Theory and Experience*, ed. Glazer and Moynihan, pp. 84-110.

central governments has led to increasing governmental attempts to intervene in regional problems, initiatives that frequently backfire when the alleviation of regional problems proves to be unsuccessful.[53] In a similar vein, Charles Ragin has suggested that growing interdependency among previously separate groups has led to competition for similar resources and rewards, thereby fostering assertions of group rights.[54] Arguing along different lines, Milton Esman proposes that the example provided by the visible anti-colonial nationalist movements in Africa and Asia helped to spur substate nationalism in Western Europe.[55]

Each of these factors has played a demonstrable role in the rise of substate nationalism in particular cases. Taken together with the ideas on economic dependency described above, they provide us with a better understanding of the kinds of factors that must be considered in our investigations of ethnicity and nationalism. A focus on the role of large-scale historical developments, however, requires significant generalization across widely different contexts and does not necessarily tell us much about the processes and structures at the community or state level that create, drive, and shape substate nationalist feeling. In response to this problem, a third conceptual approach to substate nationalism seeks to identify and understand the role of particular internal social, institutional, or behavioral processes and structures.

Anthony Smith's study of elite behavior and the works of Palley and Gladdish on legislation and government policy are representative of this third approach. In a wide-ranging discussion of nationalism and the intelligentsia, Smith persuasively demonstrates the overwhelming role of intellectual elites in giving political form and organization to ethnic movements.[56] Empirically, it is difficult to think of any exception to this pattern. In a different vein, Palley and Gladdish both focus on the role of legislation in shaping the nature and form of substate nationalism, and they propose classification schemes based on the governmental objectives

---

[52] Arend Lijphart, "Political Theories and the Explanation of Ethnic Conflict," pp. 55-57. A similar point is made in L. J. Evenden, *Cultural Discord in the Modern World*, ed. L. J. Evenden and F. F. Cunningham (Vancouver: British Columbia Geographical Series 20, 1973), pp. 8-10.

[53] Lijphart, "Political Theories and the Explanation of Ethnic Conflict," pp. 55-57. See also, Esman, "Perspectives on Ethnic Conflict in Industrial Societies," p. 388.

[54] Charles C. Ragin, "Ethnic Political Mobilization: The Welsh Case," *American Sociological Review*, 44, 4 (August 1979), pp. 619-634.

[55] Esman, "Perspectives on Ethnic Conflict in Industrial Societies," p. 376.

[56] Anthony Smith, "Nationalism, Ethnic Separatism and the Intelligentsia," pp. 17-41. In the same study Smith adopts the approach of those focusing on the role of major historical developments, arguing that rapid urbanization promoted ethnic consciousness by bringing groups into closer proximity to one another and fostering intergoup competition.

embodied in legislative approaches to ethnonationalist issues.[57] Although other examples might be cited, the important point to emphasize is that studies such as these are attempts to build theory by identifying and explaining the role of particular internal processes and arrangements in the rise of substate nationalism. They provide insight into the internal logic of substate nationalism and the kinds of questions that need to be asked when analyzing particular situations. They do not, however, necessarily reveal the interrelationships among factors or provide an overall explanation of the forces behind the rise of substate nationalist movements.

There is, of course, considerable overlap in the approaches outlined above, but categorizing theoretical contributions in terms of these perspectives provides a means of orienting work on issues of ethnicity and nationalism. Important insights have come from each of these perspectives, but each has certain weaknesses. Growing awareness of the limitations of theoretical perspectives that vest primary explanatory power in a single social structure or historical development has meant that some of the most important contributions in recent years have come from those seeking to identify and explain the constellation of relevant internal social, institutional, and behavioral processes and structures that reflect and shape ethnonationalism. Indeed, contributions of this sort are a necessary prerequisite to the construction, refinement, and assessment of grand theories of social change as applied to substate nationalist issues.[58] In addition, they provide insight into the kinds of questions that need to be asked in ongoing efforts to cope with ethnonationalist conflict. Yet identification and examination of many internal processes and structures are not well developed.

The present study arises out of the lacunae in the literature on substate nationalism regarding the processes and structures related to territory and to the conceptual, functional, and formal division of space. Territoriality (a behavioral process) and the formation of regions (social/political structures) are of profound importance in the rise of

---

[57] Claire Palley, *Constitutional Law and Minorities* (London: Minority Rights Group Report No. 36, 1979); Kenneth R. Gladdish, "The Political Dynamics of Cultural Minorities," in *The Future of Cultural Minorities*, ed. Anthony E. Alcock, Brian K. Taylor and John M. Welton (New York: St. Martin's Press, 1979). The classification schemes have been criticized, however, for their failure to take into account the different contexts within which laws are adopted and their assumption that states adopt essentially homogeneous policies toward minority ethnic groups. Alexander B. Murphy, "Partitioning as a Response to Cultural Conflict," *Geographical Perspectives*, 5/*Great Plains-Rocky Mountain Geographical Journal*, 13 (Spring 1985), p. 55.

[58] Ideas about the role of intellectual elites, for example, are integral to the conceptualization and understanding of such larger processes as capitalist development or center-periphery relations in the rise of substate nationalism.

substate nationalism. In view of the scant attention these issues have received in the literature, an explicit examination of their significance for the rise of substate nationalism is required.

Territoriality and Regionalism

The achievement of nationalist objectives is inextricably tied to issues of territorial control. In fact, definitions of nationalism usually include references to its territorial dimension. Fairchild, in his *Dictionary of Sociology*, defines a nation as "a nationality that has achieved the final stage of unification represented by its own political structure and territorial establishment."[59] Symmons-Symonolewicz's definition of nationalism discussed above also makes specific reference to territory, describing a nation as "a territorially-based community of human beings."[60] The concepts of territory and nationalism are inseparable, both because the historical association between people and place is at the heart of ethnic identity,[61] and because nationalism is fundamentally tied to struggles for control over land.[62] Moreover, most of the significant ethnic tensions in the modern world arise between groups associated, by themselves and by others, with distinctive regions within states. In fact, ethnic groups are frequently defined in regional terms because territorial considerations are so fundamental to the maintenance of group boundaries and, by extension, to ethnic group survival.

Territory is sometimes thought of in purely physical terms, representing a formal or informal spatial compartment of the earth's surface. As Knight has noted, however, territory in the nationalist context is "space to which identity is attached by a distinctive group who hold or covet that territory and who desire to have full control over it for the group's benefit."[63] Knight's approach highlights the psychological dimension of territory, mirrored in Gottmann's contention that territory is best understood as an expression of the ideas of human groups.[64] As such,

---

[59] Henry Pratt Fairchild, *Dictionary of Sociology* (New York: Philosophical Library, 1944), p. 201.

[60] Symmons-Symonolewicz, "The Concept of Nationhood," p. 221.

[61] See generally David B. Knight, "Canada in Crisis: The Power of Regionalisms," in *Tension Areas of the World*, ed. D. B. Bennett (Champaign, Illinois: Park Press, 1982), pp. 254-279.

[62] Colin Williams and Anthony D. Smith, "The National Construction of Social Space," *Progress in Human Geography*, 7, 4 (1983), p. 502.

[63] David B. Knight, "Identity and Territory: Geographical Perspectives on Nationalism and Regionalism," *Annals of the Association of American Geographers*, 72, 4 (1982), p. 526.

"territory is not; it becomes, for . . . it is human beliefs and actions which give territory meaning."[65]

Despite the ideological basis of territory and its apparent significance for ethnic relations and aspirations, territorial developments are rarely accorded serious consideration in discussions of substate nationalism.[66] Those studies that do make reference to territorial divisions frequently treat them as unproblematic reflections of the distribution of ethnic groups. Even within the discipline of geography, few have been sufficiently concerned with the causes and processes of ethnonationalism to move beyond the descriptions of ethnic patterns and traits that form a part of most regional geographies or the cursory examination of ethnic issues that appear in many political geography textbooks.[67] Nonetheless, the geographer's concern with spatial issues and the relationships between people and place has been the basis for some important contributions.

Geographers interested in issues of ethnicity and nationalism played an important role when the political map of Europe was being redrawn immediately after World War I. In 1917 Leon Dominion published a major work on the distribution of linguistic and national groups in Europe,[68] and Isaiah Bowman, among others, brought a geographer's knowledge of these matters to the Paris Peace Conference.[69] These geographers and their colleagues were fundamentally concerned with the nation-state concept as embodied in the ideal of coextensive political and cultural territories, and they encouraged thinking about politically organized areas in ethnic or cultural terms. Their conception of territory was, however, largely tied to distributional considerations.

Considerably later, a few geographers began to focus in more detail on the nature and consequences of ethnic diversity, and particularly on

---

[64] Jean Gottmann, *The Significance of Territory* (Charlottesville: University of Virginia Press, 1973), p. 15.

[65] Knight, "Identity and Territory," p. 517.

[66] Williams and Smith stated the case even more strongly when they wrote that "despite the fact that both land and space have become political territories and national homelands, geographers and sociologists alike have consistently neglected these political dimensions of 'land' and 'territory' in their treatment of environment and society." Williams and Smith, "The National Construction of Social Space," p. 502.

[67] Knight, "Identity and Territory," p. 518.

[68] Leon Dominion, *The Frontiers of Language and Nationality in Europe* (New York: American Geographical Society, 1917).

[69] Charles Seymour, *Geography, Justice, and Politics at the Paris Conference of 1919* (New York: American Geographical Society, 1951).

functional considerations.[70] The new approach represented an important shift away from viewing territory in terms of strictly distributional issues to a concern with the existence and impact of areal differentiation in ethnic patterns and processes. Moreover, geographers focusing on the nature of the state were beginning to ask questions about the significance of territory. Gottmann, following up on Hartshorne's notion of a state idea (raison d'être), argued for the importance to state integrity of iconography, by which he meant the common memories, symbols, and experiences of a people, including their identification with a particular place.[71] Jones addressed the issue of political territoriality more directly, proposing a chain of processes through which political aspirations acquire territorial expression.[72] What emerged from all these contributions was an understanding that territorial issues are inextricably tied to processes of state formation and function.

More recently, a few geographers have begun to theorize more directly about territorial processes as they relate to nationalist issues. Whebell has proposed a model of territorial separatism;[73] Knight has explored the relationships between ethnic identity and territory;[74] Williams, together with the sociologist Smith, has charted various physical and ideological dimensions of "national territory";[75] and Soja has discussed the role of territoriality in the rise of nationalism.[76] Moreover, ideas about territorial differentiation are implicit in the theories of social scientists from a variety of disciplines on core-periphery relations, internal colonialism, and uneven development. Despite the advances represented by these

---

[70] For example, Norton S. Ginsburg and Chester F. Roberts, Jr., *Malaya* (Seattle: University of Washington Press, 1958).

[71] Jean Gottmann, "The Political Partitioning of Our World," and "Geography and International Relations," *World Politics*, 3, 2 (1951), pp. 153-173. For a statement of Hartshorne's ideas, see Richard Hartshorne, "The Functional Approach in Political Geography," *Annals of the Association of American Geographers*, 60, 2 (1950), pp. 95-130.

[72] Stephen Jones, "A Unified Field Theory of Political Geography," *Annals of the Association of American Geographers*, 64, 2 (1954), pp. 111-123.

[73] C. F. J. Whebell, "A Model of Territorial Separatism," *Proceedings of the Association of American Geographers*, 5 (1973), pp. 295-298.

[74] Knight, "Identity and Territory," pp. 514-531.

[75] Williams and Smith, "The National Construction of Social Space," pp. 502-518.

[76] Edward W. Soja, *The Political Organization of Space*, Commission on College Geography Resource Paper No. 8 (Washington, D.C.: Association of American Geographers, 1971). The concept of territoriality will be discussed in more detail below.

contributions, however, research on the relationship between territory and nationalism is not well developed.[77]

Growing awareness of the interactive character of society and spatial structure suggests the importance of this line of inquiry. The works of Pred and Soja, among others, have been premised on the conceptual inseparability of social and spatial processes.[78] This perspective is receiving increasing attention outside of geography, with scholars such as Giddens urging that spatial relations be regarded as a fundamental building block of social theory.[79] Proceeding from a similar orientation, Gregory has argued that spatial structures not only are created by social processes, but themselves are agents in shaping those processes.[80] Against this backdrop, the case for an explicit focus on the territorial dimension of substate nationalism is all the more compelling.

In taking up this challenge, we must move beyond static conceptions of territory based solely on the distribution of particular phenomena to an understanding of territorial developments as processes that reflect and shape social arrangements. Theoretical frameworks such as internal colonialism and uneven development fail to meet this challenge. They imply territorial distinctions based on economic criteria, but they tell us very little about the processes linking individual and group actions with the acceptance and institutionalization of territorial divisions, the ways in which areas that do not share the same economic or cultural characteristics are conceptually and formally lumped together within regional constructs, or the role of conceptual and formal compartmentalizations of space in changing social patterns and processes. At the heart of these issues are the concepts of territoriality and regionalism.

---

[77] Thus, Colin Williams recently noted that "[d]espite earlier work on culture areas, national frontiers and homeland development, few geographers have addressed themselves to an explicit analysis of the way in which political leaders have identified and channeled group expectations to territorial control, a fascinating and worthy project to study." Colin H. Williams, "Conceived in Bondage--Called unto Liberty: Reflections on Nationalism," *Progress in Human Geography*, 9, 3 (September 1985), p. 340.

[78] See, for example, Allan Pred, "Place as Historically Contingent Process: Structuration and the Time-Geography of Becoming Places," *Annals of the Association of American Geography*, 74, 2 (1984), pp. 279-297; Edward W. Soja, "Regions in Context: Spatiality, Periodicity, and the Historical Geography of the Regional Question," *Environment and Planning D: Society and Space*, 3 (1985), pp. 175-190.

[79] Anthony Giddens, *The Constitution of Society* (Berkeley: University of California Press, 1984), p. 365.

[80] Derek Gregory, *Ideology, Science and Human Geography* (New York: St. Martin's Press, 1978). See also, Soja, "Regions in Context," p. 177.

Robert Sack defines territoriality as "the attempt to affect, influence, or control actions and interactions (of people, things, and relationships, etc.) by asserting and attempting to enforce control over a specific geographical area."[81] Defined in these terms, territoriality is explicitly a subject for geographical inquiry, but many geographers have occupied themselves with questions of spatial contact at the expense of issues relating to human strategies toward space. As Sack has put it, "to ignore territoriality or simply to assume it as part of the context is to leave unexamined many of the forces molding human spatial organization."[82]

Soja is one of the few geographers to address specifically the relation between territoriality and nationalism.[83] For Soja, territoriality grows out of physical proximity, a similarity in social, cultural, economic, and political attributes, and the consequences of functional interdependence. He posits a transformation from social definitions of territory in kinship societies to territorial definitions of society in the era of the nation-state. Human territoriality in the modern world is argued to be a fundamental link between the political system and the geographical context. Soja points out that territoriality encompasses attitudes toward place and space at a variety of levels from the individual, to the group, to the society as a whole. It is associated with a sense of place, feelings of exclusiveness, and the compartmentalization of human activities.

The concept of territoriality is significant because it "provides an essential link between society and the space it occupies."[84] Sack has expanded on this idea by seeking to identify the functions and consequences of territorial behavior.[85] Briefly, he sees territoriality as: (1) an efficient form of classification; (2) an easy mode of communication; (3) an effective strategy for enforcing control; (4) a means of reifying power; (5) a method for

---

[81] Robert D. Sack, "Territorial Bases of Power," in *Political Studies from Spatial Perspectives*, ed. Burnett and Taylor, p. 55. The adoption of Sack's definition carries with it a rejection of an understanding of territoriality based on analogies to the behavior of animals. Cf. Robert Ardrey, *The Territorial Imperative* (New York: Atheneum, 1966). The debate over the relationships between human and animal territoriality is discussed in John R. Gold, "Territoriality and Human Spatial Behavior," *Progress in Human Geography*, 6, 1 (March 1982), pp. 44-49; Torsten Malmberg, *Human Territoriality* (New York: Mouton, 1980).

[82] Robert D. Sack, "Human Territoriality: A Theory," *Annals of the Association of American Geographers*, 73, 1 (1983), p. 55.

[83] Soja, *The Political Organization of Space*, pp. 19-39. He defines territoriality as "a behavioral phenomenon associated with the organization of space into spheres of influence or clearly demarcated territories which are made distinctive and considered at least partially exclusive by their occupants or definers." *Ibid.*, p. 19.

[84] *Ibid.*, p. 33.

[85] Sack, "Human Territoriality: A Theory," pp. 58-59.

diverting attention away from other relationships; (6) a way of making relationships impersonal; (7) a seemingly neutral approach to making a place or clearing a space for things to exist; (8) a device that promotes spatial compartmentalization; (9) a way of altering the relationships between places and things; and (10) a means of promoting additional territorial arrangements. Sack further examines how these tendencies of territoriality can combine to produce different socio-spatial outcomes. Sack's work is important because it draws into the territorial equation issues of attitudes, power, and conflict, and because it emphasizes the functions and consequences of human perceptions of space and place. As will become clear in the course of this study, Sack's ideas are suggestive of a number of the factors involved in the growth, acceptance, and institutionalization of linguistic regionalism in Belgium.

It has already been pointed out that nationalism is inextricably tied to a struggle for control over land. As a result, an examination of how territory is conceptualized and used, how it gains informal and formal significance, and the consequences of these processes, is fundamental to an understanding of nationalist struggles. The ideas of Soja and Sack on territoriality are of obvious importance in this regard because they focus attention on the processes of spatial organization. As individuals, groups, and institutions act territorially in ethnically divided states, territory can become conceptually, functionally, and formally compartmentalized, creating discontinuities in patterns of spatial activity and restructuring human interaction. The resulting spatial units are regions of ethnic or national significance.

Regions, as used in this context, are created and defined by territorial behavior. Looking at regions in this way avoids the conceptual pitfalls of assuming that regional divisions are preordained, unproblematic spatial units or of separating regional constructs from social relations.[86] Regions are the product of regionalism, which Markusen defines as "the espousal of a territorial claim by some social group."[87] At the same time, the creation of regions serves to encourage regionalism. To be more specific with respect to issues of scale and scope, a region in the substate nationalist context should be understood as an informal or formal unit of territory resulting from the territorial behavior of ethnic groups (or other actors in response to ethnic considerations) with a degree of conceptual, functional, or political unity

---

[86] See generally, Doreen Massey, "Regionalism: Some Current Issues," *Capital and Class*, 6 (Autumn 1978), pp. 106-125.

[87] Ann R. Markusen, "Regions and Regionalism," in *Regional Analysis and the New International Division of Labor: Applications of a Political Economy Approach*, ed. Frank Moulaert and Patricia Wilson Salinas (Boston: Kluwer & Nijhoff, 1983), p. 35.

that serves to distinguish it from other areas. Regionalism, then, is the process by which regions are created, and the result of their creation.

The fact that people sharing a cultural or racial trait live in a particular area does not tell us much. How they and others see that area, and how those visions change, however, may tell us a great deal. Focusing on the concepts of territoriality and regionalism in the substate nationalist context directs attention to these issues by raising questions about the ways in which people and institutions attach significance to place and the implications of those attachments. At the same time, it underscores the point that regions are not just inevitable spatial manifestations of the distribution of certain phenomena, but are social creations arising out of particular historical and geographical contexts. In short, territoriality and regionalism focus attention on the importance of territorial developments and their implications for the rise of substate nationalism.

Regional developments are of significance in the substate nationalist context because they reflect and affect the territorial basis for nationalist movements, the diffusion of nationalist ideas, senses of territory, the nature and type of disputed issues, formal and informal efforts at conflict resolution, and the interrelationships among conceptual, functional, and formal compartmentalizations of space. The significance of regional developments is suggested by the contrasts between Belgium's and Switzerland's experiences with ethnolinguistic conflict.

Belgium developed as a unitary state initially with no significant regional divisions corresponding to patterns of language usage. During the twentieth century, political/cultural regions were grafted onto this structure, accompanied by significant tensions and strife. Switzerland, by contrast, developed as a confederation in which the cultural boundaries among linguistic groups largely coincided with the borders between the semi-autonomous cantons. Baetens Beardsmore explains the comparative lack of prominent intergroup conflict in Switzerland as follows:

> In general the political dimension of language contact is absent in Switzerland since, although one is also confronted with several linguistic communities within one nation state, these communities tend less to consider each other as potentially threatening to their identity. In part this can be attributed to the federal composition of the Swiss constitution where regional linguistic identity tends to coincide with federal boundaries. Moreover, mixed bilingual areas requiring protective legislation for minorities are less prevalent since each linguistically determined federal component is matched by geographic delimitations of a natural origin. Historically most of the federal states have been monoglot in composition for centuries so

that there has been less cause for friction due to language loyalties conflicting with political allegiances.88

In other words, the differences in the evolution of regional structures in Switzerland and Belgium are at the heart of the matter. With linguistic regions of considerable autonomy established from the founding of the Swiss state, Luethy could comment that "Switzerland, in the main, never resolved the political problem of multilingualism, *it avoided posing it,* which is something else altogether."89 The recent conflict in the Jura canton of Switzerland serves only to strengthen the argument since the Jura was one of the few exceptions to the spatial coincidence of linguistic and political boundaries.

The obvious importance of changing territorial conceptions for substate nationalism suggests the need for a better understanding of the nature, evolution, and implications of ethnoregionalism. Although much work has been done to identify the physical and distributional aspects of regions and to describe and analyze their functional significance, we still know very little about the processes of regional creation or the implications of those processes. The evolution of linguistic regionalism in Belgium provides a useful backdrop for investigating these issues in a particular context.

Linguistic Regionalism in Belgium

The story of the rise of ethnolinguistic consciousness and substate nationalism in Belgium has been told many times from a wide variety of perspectives. In most accounts, mention is made of the unitary character of the Belgian state in 1830, the adoption by leaders of the Flemish movement of a territorial approach to linguistic grievances at the beginning of the twentieth century, and the formal establishment of politically significant language regions during the past twenty-five years. Rarely, however, are these developments analyzed in detail in their own right. Rather, they are usually treated unproblematically. Even more troublesome is the previously noted tendency of many commentators to extrapolate from the prominence of the present regional divisions in Belgium an assumption that Wallonia and Flanders can be employed as spatial units in regional

---

[88] Hugo Baetens Beardsmore, "Linguistic Accommodation in Belgium," *Brussels Pre-Prints in Linguistics*, 5 (March 1981), p. 19. Baetens Beardsmore added one other non-territorial explanation, the comparative equality in international standing of the three main languages of Switzerland as opposed to the two main languages of Belgium.

[89] Herbert Luethy, "Incidences politiques de la pluralité des langues en Suisse," *Schweizer Monatshefte*, 46 (August 1966), p. 403. (Translation by author.)

analyses long before they acquired conceptual, functional, or formal significance.

This study takes a different perspective, seeking to address directly the nature and implications of evolving linguistic regionalism. In documenting the rise of substate nationalism in Belgium, no attempt is made to provide an exhaustive account of what happened when or to summarize the vast historical literature on the subject. Rather, the goal is to document and understand the most important developments in the process by which language areas acquired conceptual, functional, and formal significance. Thus, the inquiry seeks to identify the major changes that have taken place in the regional structure of Belgium, and to explore the reasons for, and consequences of, these particular changes.

The concepts of territoriality and regionalism are central to the analysis because the linguistic regions of Belgium, as social creations, are manifestations of attempts to assert and enforce control over territory for economic, political, or cultural ends. As the major changes in the regionalization of Belgium are presented in this study, an effort is made to identify the principal territorial actors, the forces promoting territorial behavior, and the sorts of consequences resulting from territorial developments. With respect to the first point, it is important to note that not only individuals and ethnic groups, but governments, act territorially. From a policy perspective, a basic dichotomy exists between executive, legislative, and judicial initiatives that apply to all persons of a certain status and those that affect only the inhabitants of a particular area.[90] McRae uses the terms personality and territoriality, respectively, to refer to these two different approaches.[91] Formal regions based on ethnic criteria within states are the result of governmental territoriality (for example, Finland, Yugoslavia), whereas personality approaches involve legislation based solely on individual characteristics (for example, the language laws of Ireland or Luxembourg).[92]

There has been little investigation of, or theorization about, the factors promoting territorial behavior in the ethnonationalist context. At the heart of the matter from the perspective of those actors with nationalist aspirations is the physical and symbolic significance of control over land for

---

[90] Rémi Rouquette, "Plurilinguisme et institutions politiques (Belgique, Canada, Luxembourg, Suisse)" (DEA de Droit Publique, Paris X Nanterre, 1982), no. 27.

[91] Kenneth D. McRae, "The Principle of Territoriality and the Principle of Personality in Multilingual States," *Linguistics*, 158 (August, 1975), pp. 33-54. McRae points out that these concepts "have rarely been analyzed in depth or studied comparatively." *Ibid.*, p. 33.

[92] See generally Guy Héraud, "Pour un droit linguistique comparé," *Revue Internationale de Droit Comparé*, 23, 2 (April-June 1971), pp. 313-314.

the establishment and maintenance of a degree of autonomy. In the words of Jean-François Cavin, "[s]ans assise territoriale, une communauté politique semble condamnée à dépérir."[93] Beyond this are issues of group cohesiveness, the protection of cultural/ideological attributes or material position, and the assertion and fulfillment of minority group rights and aspirations in the context of a disadvantageous political or social tradition. These points will be developed in the discussion of particular territorial initiatives in Belgium, but a few additional words need to be said about this last point.

Although the liberal-democratic political traditions of the West give ethnic groups considerable latitude to organize and press their claims, the emphasis on the individual poses problems for the presentation and defense of group rights. In fact, Enlightenment thinkers such as Locke, Rousseau, and Mill assumed that the state would be essentially ethnically homogeneous, reflecting the problems ethnic segmentation poses for the liberal conception of society.[94] Indeed, individually based democratic principles can work to the disadvantage of an ethnic minority seeking recognition within the state by discouraging access to and control over political power on a group basis.[95] At the same time, individual equality of treatment accelerates the homogenization process by imposing uniform standards of judgment on everyone.[96] In light of these factors, it is hardly surprising that ethnic groups may seek a basis for recognition and participation outside of the individualistic framework of the liberal-democratic tradition. The principle of territoriality provides an attractive alternative.

Turning to the dynamics of conflict resolution at the state level, Lijphart has proposed a now widely accepted model known as *consociational democracy* to explain conflict management in states such as Belgium and Switzerland.[97] According to Lijphart, stability is created in

---

[93] Jean-François Cavin, *Territorialité, nationalité et droit politique* (Lausanne: Held, 1971), p. 69. (". . . without a territorial foundation, a political community seems destined to decline.")

[94] Van Dyke, "The Individual, the State and Ethnic Communities," pp. 726-727.

[95] See generally, Uri Ra'anan, ed., *Ethnic Resurgence in Modern Democratic States: A Multidisciplinary Approach to Human Resources and Conflict* (Elmsford, New York: Pergamon Press, 1980), p. 20.

[96] Lawrence J. Sharpe, "Decentralist Trends in Western Democracies: A First Appraisal," in *Decentralist Trends in Western Democracies*, ed. Lawrence J. Sharpe (Beverley Hills: Sage, 1979), pp. 59-62.

[97] Arend Lijphart, *Democracy in Plural Societies: A Comparative Exploration* (New Haven: Yale University Press, 1977).

small plural states through joint governance by an elite cartel representing each of the major constituent groups within the state. Leaders negotiate on behalf of their constituents, but are committed through a complex set of administrative mechanisms to maintaining the integrity of the polity. While Lijphart's model is suggestive of certain important functional aspects of government in plural states, Covell, Claeys, and others have criticized it for its emphasis on consensual cooperation at the expense of analyzing the nature and context of actual controversies and processes of accommodation.[98] In fact, the consociational model tells us little about the reasons for the adoption by governments of territorial approaches to conflict resolution, since territorial solutions grow out of particular historical and geographical contexts and are usually invoked only in circumstances of prolonged and significant struggle.[99] Since they typically involve the devolution of at least some power from the center, they are likely to be adopted only in the face of substantial tension and pressure.[100]

To confront this issue, a more explicit focus on the contexts and processes of conflict regulation is required. Nordlinger's work on conflict regulation is important in this regard. He points to four principal factors motivating conflict regulators: (1) a desire to ward off pressure from other states; (2) a desire to avoid the negative economic consequences of conflict; (3) a desire to retain governmental power; and (4) a desire to avoid bloodshed within the regulator's own group.[101] These factors are helpful in understanding conflict regulation in Belgium. An explanation for the adoption of a territorial approach, as opposed to one based on personality, to resolve conflicts, however, requires a consideration of other factors. Dikshit has pointed to the importance of independent or competitive economies among ethnic groups, a prior political status for the area in question, the absence of a common external threat, and the presence of incompatible

---

[98] Maureen Covell, "Ethnic Conflict and Elite Bargaining: The Case of Belgium," *West European Politics*, 4, 3 (October 1981), pp. 197-218; Paul H. Claeys, "Political Pluralism and Linguistic Cleavage: The Belgian Case," in *Three Faces of Pluralism*, ed. Stanislaw Ehrlich and Graham Wooton (Westmead, England: Gower, 1980), pp. 171-172.

[99] Murphy, "Partitioning as a Response to Cultural Conflict," pp. 56-58.

[100] See generally, Donald N. MacIver, "Ethnic Identity and the Modern State," in *National Separatism*, ed. Colin H. Williams (Vancouver: University of British Columbia Press, 1982), pp. 299-307.

[101] Eric Nordlinger, *Conflict Regulation in Divided Societies*, Harvard Studies in International Affairs, #29 (Cambridge, Massachusetts: Center for International Affairs, Harvard University, 1972), pp. 43-52. He argues that these motives are associated with four structural conditions: (1) a low position of the state in terms of international power, making it subject to external pressure; (2) the existence of a substantial commercial class with expansionist economic interests; (3) a stable balance of power in which no party can be entirely dominant; and (4) a belief that continued conflict will lead to suffering.

social, economic, and political ideologies.[102] As will become clear in the course of this study, some of these factors are relevant in the Belgian case, but of equal importance are distributional considerations, the nature of the particular issues posed by the competing ethnic groups, the degree of prior social and cultural division along ethnoregional lines, and the perceptions by decision makers of the implications of territorial arrangements for their own control of power within the state.

Turning to the implications of territorial developments, a central concern of this study is the interrelationships among conceptual, functional, and formal regional divisions. An effort is made to explore how compartmentalizations of space of one type reflect and affect spatial divisions of other types. Important to this approach is the idea that the relation between conceptual or functional regions and formal regions is not unilinear. Rather, formal regions have a profound impact on both conceptual and functional spatial divisions, and vice versa.[103] The objective of this study, then, is to identify the more important features of the regional dynamics of language differentiation in Belgium in an effort to gain a better understanding of the significance of territorial developments in the substate nationalist context.

The evolution and implications of linguistic regionalism in Belgium are traced and analyzed through a focus on the discourse of ethnic group leaders and legislators, and through the changing framework of language legislation. With regard to the former point, Verdoodt has emphasized the difficulty of a careful and balanced analysis of Belgium's language problems in view of the substantial bias of most studies.[104] These same studies, however, provide valuable clues to the perspectives and attitudes of participants in language controversies throughout Belgium's history. Focusing on the concerns expressed by the authors rather than their descriptions of events provides a means of gaining some understanding of evolving regional conceptions. Among the most important sources for this kind of analysis are documents of the ethnonationalist movements, the writings of ethnic group leaders, reports of commissions studying the language situation, and transcripts of legislative debates on language questions.

---

[102] Ramesh D. Dikshit, *The Political Geography of Federalism* (New Delhi: Macmillan of India, 1975), p. 227.

[103] Soja, *The Political Organization of Space*, p. 4.

[104] Albert Verdoodt, *Les problèmes des groupes linguistiques en Belgique*, Bibliothèque des Cahiers de l'Institut de Linguistique de Louvain, 10 (Louvain: Editions Peeters, 1977), preface.

Belgium probably has one of the most complex sets of laws relating to internal ethnic differences of any state in the world. The significance of law as an instrument of social change is a well-developed theme in the literature on law and society.[105] In the West, laws are the primary vehicle for maintaining social stability, regulating conflict, and legitimizing power structures.[106] As such, laws both reflect and structure social patterns and behavior. Moreover, since the state has a strong interest in promoting respect for the legal system as a forum for the resolution of conflicts, the enactment of laws is accompanied by powerful institutional mechanisms to ensure compliance. Hence, laws not only reflect social patterns and behavior, but are fundamental forces in bringing about change.

Laws are of particular importance in the substate nationalist context because the regulation of ethnic discord in the modern world has increasingly been approached through the legislative process. In fact, much of the effort of ethnonationalist groups in many states revolves around attempts to secure rights and privileges through legislative enactments. Successful efforts often have involved the granting of positive rights to ethnic groups.[107] Positive rights can take a variety of forms ranging from proportional electoral laws, to formal power-sharing arrangements, to the granting of rights based on ethnically defined territories. In the last case, the resulting laws reflect ethnic patterns and conceptions of territory as well as the ideological and political priorities of those in power. At the same time, they serve to alter the structure of political space within the state, with all that that implies in terms of cognitive and functional change.

Given the important role of law in the evolution of ethnonationalism, a focus on the legislative framework in Belgium provides useful insights into regional change. The timing and nature of the language laws of Belgium evince periods of change in conceptualizations of the regional structure of Belgium, the relative importance of territorial issues, and the motives of conflict regulators. At the same time, the territorially based language laws provide us with a frame of reference for assessing the impacts

---

[105] Many of the important themes are explored in Lee S. Weinberg and Judith W. Weinberg, *Law and Society* (Washington, D.C.: University Press of America, 1980). See also Patrick S. Atiyah, *Law and Modern Society* (Oxford: Oxford University Press, 1983).

[106] See generally, Gordon L. Clark and Michael Dear, *State Apparatus: Structures and Language of Legitimacy* (Boston: Allen & Unwin, 1984), ch. 6.

[107] "Positive rights" are those rights that are explicitly designed to preserve or promote ethnic distinctions. This is in contrast to "negative rights," which require universal individual equality of treatment. Robert G. Wirsing, "Dimensions of Minority Protection," in *Protection of Ethnic Minorities: Comparative Perspectives*, ed. Robert G. Wirsing (New York: Pergamopn Press, 1981), pp. 9-10.

of formal regionalization on social, political, economic, and cultural patterns and processes. Consequently, the statutes of Belgium, together with the documents containing the legislative history of the language laws, are important sources for this inquiry.

Since the primary focus of this study is the nature and role of changing linguistic regionalism within the sovereign Belgian state, a detailed analysis of territorial developments prior to 1830 is beyond the scope of the inquiry. Nevertheless, to provide the necessary background for the ensuing analysis, it is important to establish the nature of regional divisions in Belgium at the time of independence.

Chapter 3

LANGUAGE AND REGIONALISM IN THE EMERGING BELGIAN STATE

The modern sovereign state of Belgium came into being in 1830 after a short revolt of the southern part of the Low Countries against the rule of the Dutch king William I of Orange-Nassau. To trace the subsequent evolution of language and regionalism in Belgium, it is necessary to examine the linguistic characteristics of the newly created Belgian state and the formal and informal territorial divisions of the time. A brief overview of historical developments prior to 1830 sets the stage for the ensuing discussion.[1]

Historical Introduction

The area we now call Belgium became a part of the Roman Empire in 57 A.D., and was subjected to strong romanizing linguistic influences for several centuries.[2] Beginning in the third century, Germanic tribes invaded the northeastern parts of the Empire and started settling there, a development that accelerated greatly in the fifth century with the withdrawal of Roman troops from the area. The invaders brought with them their own language and customs, thereby changing the linguistic character of the region.

The reasons for the emergence of a rather distinct language boundary across the middle of present-day Belgium are obscure. Among the many

---

[1] Among the better known historical works covering Belgian history prior to 1830 are Henri Pirenne, *Histoire de Belgique*, 7 vols. (Brussels: Lamartin, 1902-1932), and Léon Van der Essen, *A Short History of Belgium* (Chicago: University of Chicago Press, 1915).

[2] See generally, L. Th. Maes and R. Van Santbergen, eds., *Documents Illustrating the History of Belgium*, vol. 1, *From Prehistoric Times to 1830*, Memo from Belgium no. 178 (Brussels: Ministry of Foreign Affairs, External Trade and Cooperation in Development, 1978), pp. 39-43.

theories that have been advanced are: (1) that the major Frankish invasion of 406 A.D. was decisive in bringing Germanic language and culture to the northern area, but that the invasion was halted by the presence of a great natural forest and/or by Roman defenses along the Cologne-Boulogne highway;[3] (2) that the language of the invading Franks rapidly became dominant in the north because that area was only sparsely settled by the indigenous population, whereas higher local population densities in the South led to the linguistic assimilation of the invaders;[4] and (3) that the linguistic situation remained unstable for centuries after the Frankish invasions, with later colonization finally resulting in a stable language boundary.[5] The sources of evidence are so varied and validation so difficult that Stengers has concluded that the debate over the origin of the language boundary is likely to remain inconclusive.[6]

It can be said with reasonable certainty, however, that by the tenth or eleventh century, the boundary between the Romance and Germanic tongues had been established in this part of Europe in more or less its present position. Importantly, this was not an outgrowth of the founding of feudal principalities in the area. In fact, three of these duchies, Flanders, Brabant, and Liège, straddled the language border, as did the bishoprics found in that area at the time.[7] Throughout the region, Latin remained the language of culture and the Church, while most ordinary people spoke local dialects. Coulon suggests that the lack of competition with respect to language use meant that there was no real language problem at the time.[8] Certainly, with the feudal organization of society and the multilingual

---

[3] Godefroid Kurth, *La frontière linguistique en Belgique et dans le nord de la France*, vol. 2 (Brussels: Académie Royale des Sciences, des Lettres et des Beaux-Arts de Belgique, 1895-1898), pp. 3-71. The location and characteristics of the forest are, however, matters of considerable dispute.

[4] J. Dhondt, "Essai sur l'origine de la frontière linguistique," *L'Antiquité Classique*, 16 (1947), pp. 261-286.

[5] Franz Petri, *Germanische Volkserbe in Wallonien und Nordfrankreich: Die Fränkische Landnahme in Frankreich und den Niederlanden und die Bildung der Westlichen Sprachgrenze*, vol. 2 (Bonn: L. Röhrscheid, 1937), pp. 941-1002.

[6] Jean Stengers, *La formation de la frontière linguistique en Belgique ou de la légitimité de l'hypothèse historique* (Brussels: Collection Latomus, 1959), pp. 48-53.

[7] Theo Luykx, *Culturhistorische Atlas van België* (Brussels: Elsevier, 1954), maps 5 & 6. Although the boundaries of the principalities shifted many times throughout the Middle Ages, they never conformed to linguistic distributions. For a detailed study of the principalities in this area during the Middle Ages, see Léon Vanderkindere, *La formation territoriale des principautés belges au Moyen Age*, 2 vols. (Brussels: H. Lamertin, 1902).

[8] Marion Coulon, *L'autonomie culturelle en Belgique* (Brussels: Fondation Charles Plisnier, 1961), p. 13.

principalities and bishoprics, there was no regionwide sense of ethnolinguistic identity.

The twelfth and thirteenth centuries saw the rise of the communes (free cities) in the northern part of present-day Belgium, with cities such as Ghent and Bruges developing as centers of trade and industry.[9] This transformation took place primarily in the Germanic-speaking areas within the principalities of Flanders and Brabant, bringing them unprecedented wealth and prestige. During the thirteenth century, however, the French language began to infiltrate into the upper levels of society in the Germanic part of the County of Flanders, as reflected in the Flemish poet Jacob van Maerlant's (ca. 1235-1300) criticisms of the stylistic qualities of the French language.[10] The spread of French was apparently quite limited, though, for when Brussels succeeded Leuven (Louvain) at the end of the thirteenth century as the political and administrative center of the Duchy of Brabant, the language of administration was *thiois*, a predecessor of modern Dutch.[11]

With the union of the Belgian principalities under the Francophone dukes of Burgundy at the beginning of the fourteenth century, the influence of French grew considerably. The language at the upper levels of administration was now French, and official acts began to be published in that language.[12] Moreover, Brussels, as the official residence and court seat of the Burgundian dukes, became the home of a small French-speaking upper class. The influence of French in Brussels during this period should not be exaggerated, however, since the city apparently remained overwhelmingly Germanic in speech throughout the Middle Ages.[13]

The area presently occupied by Belgium, the Netherlands, and Luxembourg, with the exception of the prince-bishopric of Liège, became a part of the Hapsburg Empire at the beginning of the sixteenth century. It was divided into seventeen provinces based on the former principalities. Figure 2 is a map showing the southern provinces in 1549 superimposed on a map of modern Belgium with the language boundary indicated. It is clear

---

[9] See generally, Léon Van der Essen, *A Short History of Belgium*, ch. 4.

[10] Discussed in Kenneth D. McRae, *Conflict and Compromise in Multilingual Societies: Belgium* (Waterloo, Ontario: Wilfrid Laurier University Press, 1986), p. 18.

[11] Marc Platel, "Bruxelles, aussi capitale des Flamands," trans. from the Dutch by Willy Devos, *Septentrion-Revue de Culture Néerlandaise*, 9, 2 (June 1980), p. 78.

[12] Coulon, *L'autonomie culturelle en Belgique*, p. 13.

[13] Inge Itta Callebaut, "The Evolution of Language and Culture: Fifty Families in Brussels (1870-1986)," paper presented at Soviet-Belgian Symposium on Multilingualism: "Aspects of Interpersonal and Intergroup Communication in Pluricultural Societies," Bibliothèque Royale Albert 1er, Brussels, March 13-15, 1986, p. 8.

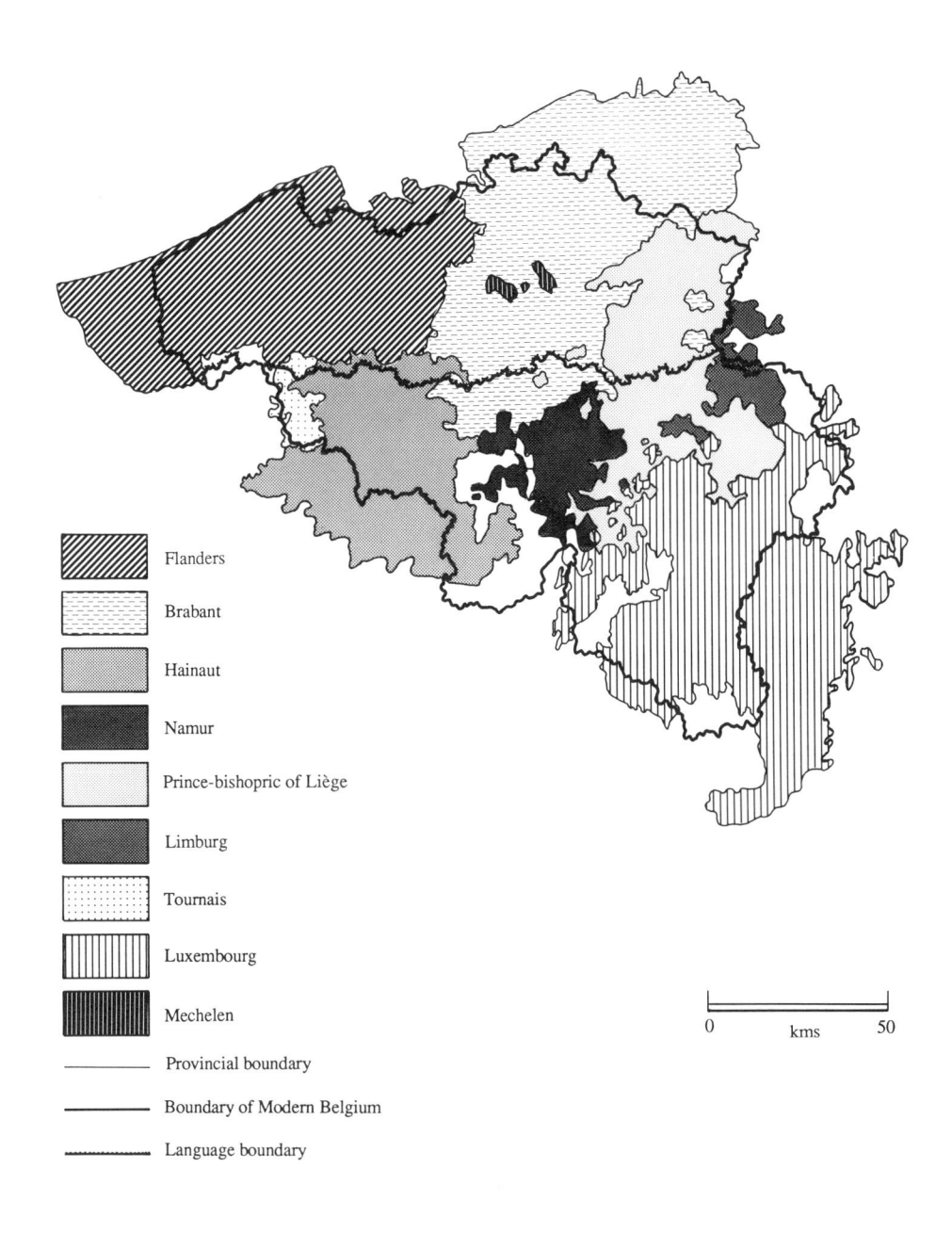

Fig. 2. Hapsburg provinces in area of modern Belgium.

from the map that there was little correspondence between the provinces and language patterns at the time. Although the beginning of the Hapsburg period is associated with a great flowering of Flemish culture, French was used to some degree in the administration at Brussels, and the emperor Charles V established a few secondary schools in the city to teach the French language.[14]

With the succession of Philip II to the Hapsburg crown in 1555, the Seventeen Provinces came to be ruled almost entirely from Spain. The despotic rule of Philip II, associated with an attempt to defend and promote Catholicism at all costs, precipitated a revolt in the Seventeen Provinces that was particularly fierce in the north where Protestantism had taken hold. In 1581, the northern provinces secured their independence, leaving the southern provinces (approximately the area now occupied by Belgium and Luxembourg) under Hapsburg control. Importantly, as a consequence of the religious wars and the establishment of the Netherlands as an independent state, many of the elite in the Germanic-speaking areas of present-day Belgium converted to Protestantism and emigrated.[15]

Although the southern provinces remained under foreign domination for more than two hundred years after the successful revolt of the Netherlands, 1581 marks the beginning of a period during which the ancestors of the modern Belgians, with the exception of those living in the prince-bishopric of Liège, came to share a political history out of which can be said to have emerged a certain sense of common historical experience and, by extension, Belgian nationalism.[16] With the exception of the period 1598-1621, when the region was under the control of Isabella, daughter of Philip II, and her husband, the archduke Albert of Austria, the provinces continued to be ruled from Spain until the beginning of the eighteenth century. They then fell under Austrian control as a result of treaties resolving the protracted struggle between the Hapsburgs and the Bourbons. The Austrian Hapsburgs retained control over the provinces until 1794, with the exception of an eight-month revolt in 1789.

Throughout the seventeenth and eighteenth centuries (i.e., during the last century of Spanish rule and the Austrian period) there was a great

---

[14] Platel, "Bruxelles, aussi capitale des Flamands," pp. 78-79. Up until this time instruction in all of the secondary schools had been carried out in Latin.

[15] Lode Claes, "Le mouvement flamand entre le politique, l'économique et le culturel," *Res Publica*, 15, 2 (1973). p. 219; Maurits Van Haegendoren, *The Flemish Movement in Belgium*, translated from the Dutch (Antwerp: Flemish Cultural Council, 1965), p. 25.

[16] Jean Stengers, "Belgian National Sentiments," in Arend Lijphart, ed., *Conflict and Coexistence in Belgium: The Dynamics of a Culturally Divided Society* (Berkeley: University of California Press, 1981), pp. 46-60.

flowering of French culture and language in Europe, a development that was to leave an imprint on the linguistic character of the future Belgian state.[17] French became the language of high society throughout much of Europe, and many of the children of the elite in the Low Countries, including the children of the Flemish bourgeoisie, were sent to Paris to study.[18] In fact, the Austrians employed French as the language of the central administration, as well as in relations between central and regional authorities. Although the impact of these measures on the linguistic character of the Flemish area is a matter of some dispute, they certainly enhanced the prestige of French and contributed to its diffusion in the Germanic-speaking areas.[19]

The growing importance of the French language was further fostered by the earlier standardization of French than of Dutch. This put French in a better position to function as a means of communication between regions.[20] In addition, it was during this period that many of the secondary schools abandoned the use of Latin and began teaching in French, even in the Flemish areas. Although this development was not unique to the Austrian Netherlands, it almost certainly had a profound impact on the linguistic evolution of the Flemish part of present-day Belgium. Not surprisingly, French had its greatest impact in urban centers such as Ghent and Brussels, where institutions of higher learning functioned in French, intellectuals were attracted by the universality of the language, and members of the nobility came into contact with frenchified Austrian nobles.[21]

The growing importance of French and the corresponding decline of variants of Dutch in the Flemish area are attested to by the publication in 1788 of a tract by the Brussels lawyer Verlooy entitled *Verhandeling op d'Onmacht der Moederlyke Tael in de Nederlanden* (Treatise on the

---

[17] See generally, Coulon, *L'autonomie culturelle en Belgique*, p. 14.

[18] Shepard B. Clough, *A History of the Flemish Movement in Belgium* (New York: Richard R. Smith, 1930), p. 16.

[19] See generally, Marcel Deneckere, *Histoire de la langue française dans les Flandres (1770-1823)*, parts II-III (Ghent: Romanica Gandensia, 1954), pp. 29-128. There is little evidence, however, that provincial or local administrative officials in the Germanic-speaking areas used French during this period.

[20] For a detailed discussion of the standardization of French and Dutch and its impact on language use see Arthur E. Curtis, "New Perspectives on the History of the Language Problem in Belgium" (Ph.D. dissertation, University of Oregon, 1971), pp. 83-180. Importantly, a standardized form of Dutch did not become accepted in the northern part of Belgium until well after the founding of the Belgian state.

[21] French never played an important role in primary education in the Flemish areas, however. It was limited to secondary schools and, somewhat later, the universities that had previously been conducted in Latin.

Powerless Position of the Mother Tongue in the Netherlands).[22] Verlooy polemicized against the growing use of French by educated Flemings, leading many future commentators to regard him as the father of the Flemish movement. At the time, however, Verlooy's plea had little effect. Rather, increasing administrative centralization by the Francophone government together with the growing dominance of French among the clergy, the nobility, the judiciary, and the scientific community signaled the growing influence of French.[23]

The victory of the French over the Austrians in the southern Low Countries in 1794 greatly enhanced the position of the French language in the region. The area of present-day Belgium, including the former prince-bishopric of Liège and the Grand Duchy of Luxembourg, was incorporated into France in 1795 and remained under French domination for twenty years. The French organized the territory into nine departments.[24] The basis for the delimitation of the departments is not well understood. The departmental boundaries corresponded in some places to the borders of the medieval provinces and in others to the administrative units established during the brief period of independence from Austria in 1789.[25] The distribution of major urban centers may have also played a role. Of particular importance for the present study is the general lack of correspondence between provincial and linguistic boundaries. Brabant straddled the Germanic-Romance linguistic divide, and the spatial conformity between the borders of several of the other provinces and the language boundary was inexact at best.[26] Moreover, the judicial districts established by the French in the region did not reflect linguistic considerations, with the Brussels district extending over most of the western part of present-day Belgium, and the Liège district covering the eastern part, with the exception of Luxembourg.[27] Even the Catholic dioceses were established without reference to linguistic considerations.

---

[22] A good biography of Verlooy is Jan Van den Broeck, *J. B. C. Verlooy Vooruitstrevend Jurist en Politicus uit de 18de Eeuw* (Amsterdam: Standaard Wetenschappelijke Uitgeverij, 1980).

[23] Deneckere, *Histoire de la langue française dans les Flandres (1770-1823)*, pt. I.

[24] The nine departments, followed in parentheses by their successor names in the Belgian state, were: Lys (West Flanders), l'Escaut (East Flanders), Deux-Nèthes (Antwerp), Meuse-Inférieure (Limburg), Dyle (Brabant), Ourthe (Liège), Jamappes (Hainaut), Sambre-et-Meuse (Namur), and Forêts (Luxembourg).

[25] See Léon Van der Essen, "La république des Etats-Belgiques-Unis 1789-1790," insert on pl. X, *Atlas de géographie historique de la Belgique* (Brussels: G. Van Oest, 1919).

[26] See generally, Theo Luykx, *Cultuurhistorische Atlas van België* , pp. 125-26; pl. 13.

[27] "La Belgique sous la domination française, 1794-1814—les départements belges et les départements voisins au 1er janvier 1812," map in the collection of the Bibliothèque Royale

The period of French domination was of particular significance for the linguistic evolution of the region because of French efforts to impose their language throughout the nine departments. The French were apparently motivated by the belief that the departments could be successfully administered as part of France only if French became the language of all of the people.[28] As a result, laws were passed requiring the use of French in administrative and judicial matters (1803), replacing Latin with French in the universities (1806), and curtailing the use of Dutch or Flemish dialects by the press (1807).[29] As a consequence of these developments, in the Flemish area upward mobility came to be tied to a knowledge of French, and a Francophone bourgeoisie emerged in most of the major urban centers. Moreover, the establishment of secondary and university education throughout the region in French ensured that French would be the language of the intelligentsia for some time to come. These measures had little impact on the peasants, however, who continued to speak Flemish dialects.

The defeat of Napoleon at Waterloo (only a few kilometers south of Brussels) marked the end of French rule in the region. At the Congress of Vienna, the area of present-day Belgium and Luxembourg was united with the Netherlands to form the Kingdom of the United Netherlands. The new state was ruled from Amsterdam by the Dutch king William I. The king initiated a linguistic policy aimed at weakening the influence of French in the Flemish areas.[30] He issued decrees making Dutch the only legally recognized language for public affairs in the Flemish areas beginning in 1823, creating a Dutch language teacher's college in Lier, mandating the use of Dutch in secondary education in the Flemish areas, and requiring that Dutch be taught as the second language in the Francophone areas.[31] Although these measures were not all successfully enforced, they did lead to the removal of some functionaries who lacked a sufficient knowledge of

---

Albert 1er, Brussels, Belgium (Brussels: . G. Van Oest, 1812). Belgium was also divided into law enforcement districts, revenue inspection districts, military districts, university districts, and mining districts during the French period without reference to linguistic considerations. Françoise de Dainville and Jean Tuland, *Atlas administratif de l'empire français*, after the atlas produced by order of the Duc de Feltre in 1812 (Paris: Librairie Minard, 1973), maps VI, XII-2, XV, XVI-3, and XVI-4.

[28] See generally, Clough, *A History of the Flemish Movement in Belgium*, pp. 20-21.

[29] See generally, Deneckere, *Histoire de la langue française dans les Flandres (1770-1823)*, pp. 131-277.

[30] André Monteyne, *De Brusselaars in een Stad die Anders is* (Tielt - Bussum: Lannoo, 1981), p. 205.

[31] For a detailed discussion of the language laws of the period see, Floris Blauwkuip, *De Taal Besluiten van Koning Willem I* (Amsterdam: De Bussy, 1920).

Dutch. In the absence of an indigenous Dutch-speaking elite, individuals from Holland were often brought in as replacements.[32] The resentment that followed transcended language lines.

King William's measures may have succeeded in halting the growth of French in the Flemish areas, but there is little evidence that they had much effect in reversing the frenchification that had already taken place. In fact, many Francophones retained their positions, and William's policies engendered considerable hostility among the French-speaking Flemish bourgeoisie. The opposition became so intense that, as will be seen later, William was eventually forced to make Dutch optional in the Flemish areas, and the language issue played an important part in bringing about the Belgian Revolution of 1830.

During the period of unification with the Netherlands a sense of Flemish cultural consciousness developed among a small group of intellectuals, who were inspired by the new ties with a Dutch-speaking intelligentsia.[33] Moreover, King William's imposition of Dutch as the language of public affairs in the Flemish-speaking areas marked the first substantial initiative of a governing authority that had differential territorial impact based on the linguistic geography of the area. For the first time, the Germanic and Romance language areas were accorded a degree of regional significance. He even introduced the idea of changing administrative borders to correspond to the language boundary, a policy that resulted in the adjustment of some district boundaries in Brabant, but went no further.[34] Taken as a whole, however, the impact of William's measures was temporally and substantively limited. Hence, the period of union with the Netherlands did not mark the forging of any significant sense of regional consciousness based on linguistic considerations.

Before discussing the Belgian Revolution, mention should be made of economic developments that were to have a profound impact on the language question in Belgium. During the early decades of the nineteenth century, the Industrial Revolution arrived in Belgium. With the notable exception of the textile industry, much of the associated development took

---

[32] Maroy, "L'évolution de la législation linguistique belge," p. 452.

[33] Manu Ruys, *The Flemings: A People on the Move, a Nation in Being*, translated from the Dutch by Henri Schoup (Tielt: Lannoo, 1981), ch. 1.

[34] Curtis, "New Perspectives on the History of the Language Problem in Belgium." According to Curtis, William did not carry this idea to the adjustment of provincial boundaries because "he found that the (Francophone) bourgeoisie in the Flemish-speaking communes of Walloon provinces wanted to remain in these provinces, while the bourgeoisie of the Walloon-speaking communes in Flemish provinces often welcomed (the idea of) a shift to a Walloon province." *Ibid.*, p. 428.

place in the Sambre-Meuse valley close to coal deposits that had been used in small metalworks in the area since the early 1500s.[35] The first coke-burning blast furnace on the European continent opened at Liège in 1823, and by the dawn of Belgian independence, the Sambre-Meuse valley was poised to become one of the major industrial districts of Europe. Most of the development was concentrated in an arc running west from Mons through Charleroi and Namur to Liège and thus lay within the French-speaking part of present-day Belgium. The economic potential of this area translated into a sizeable concentration of power in the hands of French speakers in the emerging Belgian state. Even in northern Belgium the traditional industries such as textiles tended to be dominated by Flemings who had adopted the use of French.

As noted previously, the linguistic policies of King William I alienated the French-speaking bourgeoisie throughout the southern provinces of the Kingdom of the United Netherlands. Religious differences between the Catholic south and the Protestant north served further to divide the kingdom. This had a linguistic dimension, with many of the Catholic clergy, including those in the Flemish area, fearing that the Dutch language would be a vehicle for the spread of Protestantism from the north. Added to these considerations were the growing concern of the emerging Francophone industrial bourgeoisie over the low Dutch tariffs, the economic upheavals associated with early industrialization, the opposition of the clergy to the establishment of state schools and state inspection of Catholic schools, the underrepresentation of the future Belgian provinces in the central government, and the insensitivity of King William to the grievances of the future Belgians.[36] As a result, a strong opposition rose up in the 1820s that eventually succeeded in pressuring William to revise his linguistic policies. Just before the Belgian Revolution, William agreed to allow freedom of language choice throughout the southern provinces. The king's actions were too late to stave off resentment that had grown up against union with the Netherlands over religious, linguistic, and administrative issues, however, and a brief revolution broke out in 1830 that led to the establishment of Belgium as an independent state.

Although the Belgian Revolution was dominated by French speakers, it was not a revolt by Francophones against those speaking Germanic tongues, nor were its supporters seeking reunification with France.[37]

---

[35] Raymond Riley, *Belgium*, Studies in Industrial Geography (Folkstone, England: Dawson, 1976), pp. 28-30.

[36] See generally, Reginald De Schryver, "The Belgian Revolution and the Emergence of Belgium's Biculturalism," in *Conflict and Coexistence in Belgium: The Dynamics of a Culturally Divided Society*, ed. Lijphart, pp. 20-21.

Rather, the revolution grew out of resentment over what was perceived as foreign domination by the Dutch. The linguistic, religious, and socioeconomic grievances described above were of undeniable importance in promoting anti-Dutch sentiment, but for the leaders of the revolt, it was also a national revolution by a people who had shared a common political history for more than 200 years as inhabitants of an area administered more or less as a distinct region with some degree of autonomy by the Spanish, Austrians, French, and Dutch. Out of this developed an antipathy for external control that far overshadowed any internal divisions.[38] Although external intervention may have ensured the survival of the nascent Belgian state, for those segments of society leading the revolt, eighteenth century ideas of nationalism were very much a consideration.

Language and Regionalism in the Newly Created Belgian State

At the time of independence, the inhabitants of Belgium did not regard themselves as members of distinct language communities.[39] Most of the people in the southern part of Belgium spoke French, which had largely replaced the Romance dialects for most general purposes. The North, by contrast, was characterized by a population speaking either Flemish dialects or Dutch, with a small but influential group in the major urban centers using French. The term Walloon appeared in a few writings as early as the late fifteenth century in reference to the peoples of the south and their Romance dialects. There was no suggestion prior to 1830, however, that the Walloons constituted a cohesive ethnic group.[40] Before the nineteenth century, the terms Flemish or Fleming referred solely to the people and Germanic dialects of medieval Flanders. Beginning in 1814, references to "the Flemish provinces" and "the Walloon provinces" appear, but even at this stage these phrases were not used in such a way as to suggest any group cohesion among Romance or Germanic-speaking peoples.[41]

---

[37] These points are developed in Jacques Logie, *1830. De la régionalisation à l'indépendence* (Gemgloux: Duculot, 1980).

[38] This point is developed in De Schryver, "The Belgian Revolution and the Emergence of Belgium's Biculturalism," pp. 22-25; Stengers, "Belgian National Sentiments," pp. 46-60.

[39] See generally, Janet Polasky, "Liberalism and Biculturalism," in *Conflict and Coexistence in Belgium: The Dynamics of a Culturally Divided Society*, ed. Lijphart, p. 34.

[40] Albert Henry, *Esquisse d'une histoire des mots Wallon et Wallonie* (Brussels: La Renaissance du Livre, 1974), pp. 35-40. The term *Wallon* is derived from the Germanic word for the romanized Celts. From a philological perspective, Walloon is one of several dialects spoken in southern Belgium, along with Picard, Liègeois, Rouchi, Gaumais, and Brabançon. Frequently, however, the term is used to refer to all of the Romance dialects of the area.

[41] Maurits Gysseling, "Vlaanderen (Etymologie en Betekenisevolutie)," in *Encyclopedie van de Vlaamse Beweging*, ed. Jozef Deleu et al. (Tielt: Lannoo, 1975), vol. 2, pp. 1906-1910.

The lack of linguistic polarization in Belgium in 1830 is indicated by the fact that people from all over the country participated in the revolution, united by a common dislike of the Dutch. Although the majority of the revolutionaries were Francophones, this was due to certain imbalances that were to play a role in the later linguistic polarization, rather than to entrenched linguistic cleavages. Indeed, as Henri Pirenne has documented, at the meeting of the first National Congress in 1830 there was more hostility between Catholics and Liberals than between Flemings and Walloons.[42] For most of the common people, neither Belgian nor Walloon/Fleming identity was significant, because group consciousness existed only at the local level.[43] For the leaders of the newly created state, however, linguistic divisions were clearly subordinate to feelings of antipathy for external domination and a sense of Belgian nationalism.

As further support for the lack of language group consciousness, there is no evidence that either the Flemish or the Walloon provinces constituted regions in any meaningful sense within the Belgian state of 1830. In chapter 2, a region was defined as an informal or formal unit of territory with a degree of conceptual, functional, or formal unity that serves to distinguish it from other areas. The Flemish and Walloon areas of Belgium lacked unity on any of these levels in 1830. To begin with, there were no formal divisions in Belgium corresponding to language distribution. Belgium was established as a strongly centralized state on the French model. The provinces, direct progeny of the French departments established in 1795, did not reflect language patterns. Flanders referred either to the area of the medieval county of Flanders or to the provinces of East and West Flanders,[44] and the toponym *Wallonia* had not yet been coined. No reference was made in the Belgian Constitution to language regions, and no distinctions were made in early legislation based on language patterns.

Nor is there any evidence that language areas in the newly independent Belgian state had any functional integrity. Although the Sambre-Meuse valley was emerging as an industrial center, it was only a part of southern Belgium, and workers were coming to the district from all over the country. In addition, Brussels, as the capital of Belgium, enjoyed strong ties with urban centers in both the Flemish and Walloon areas.

---

[42] Pirenne, *Histoire de Belgique*, vol. 6, p. 440.

[43] This was in part a reflection of the continued importance of the agrarian village unit, a point that will be taken up in more detail below.

[44] For example, a book published in 1830 under the title *L'agriculture pratique de la Flandre* ("Agricultural Practices in Flanders") makes reference solely to the area of the provinces of East and West Flanders. J -L. van Aelbroeck, *L'agriculture pratique de la Flandre* (Paris: Madame Huzard, 1830).

Perhaps most importantly, the basic unit of economic and social organization in Belgium in 1830 was the commune. Communes were units of political organization of relatively small extent which were based on the medieval agrarian village unit. Their importance clearly overshadowed any possible functional divisions based on language distribution.[45]

The lack of formal or functional regionalization along language lines in 1830 reflects the absence of a conceptual partitioning of Belgium into linguistic regions. Territorial loyalties were either local (at the village or commune level) or, in the case of a political and social elite, national in the sense of Belgium-wide nationalism. The Germanic and Romance areas of Belgium had never been politically organized as formal units, so there was no historical basis for dividing the country into language regions. In addition, the only antecedent policy distinction that had been made between northern and southern Belgium came from the ultimately ineffective language decrees of William I during the fifteen years prior to Belgian independence, hardly the basis for the development of a strong sense of linguistic territorial differentiation. Moreover, although much has been made in the Swiss context of the role of physiographic regions in promoting a sense of cultural grouping, no such similar argument can be made for Belgium.[46] As figure 3 demonstrates, in Belgium there is little relationship between the major physiographic regions and the distribution of languages. Finally, the lack of linguistic standardization in the Flemish region, and even to some degree in the Walloon region, rendered the formation of conceptions of territorial unity based on shared linguistic attributes improbable.

In short, the inhabitants of the newly independent Belgian state did not define themselves in terms of language communities, and the linguistic geography of the country was not the basis for conceptual, functional, or formal territorial divisions at the time. This is not to suggest that there were no distinctions between northern and southern Belgium that were significant in the development of linguistic rifts. As we will see in the next chapter, these differences included the greater differences among dialects in the North, the concentration of heavy industry in part of the South, the stronger Catholic influence in the North, the existence of a French-speaking elite in the northern cities, and the closer ties of the elite in some of the southern cities with Paris than of their counterparts in the northern cities with Amsterdam. The important point to note is that at the time of

---

[45] The role of the commune in Belgium is discussed in H. Van der Haegen, M. Pattyn, and C. Cardyn, "The Belgian Settlement System," *Acta Geographica Lovaniensia*, 22 (1983), p. 255.

[46] Cf. Robert Sevrin, *Géographie de la Belgique et des Pays-Bas* (Paris: Presses Universitaires de France, 1969), p. 57.

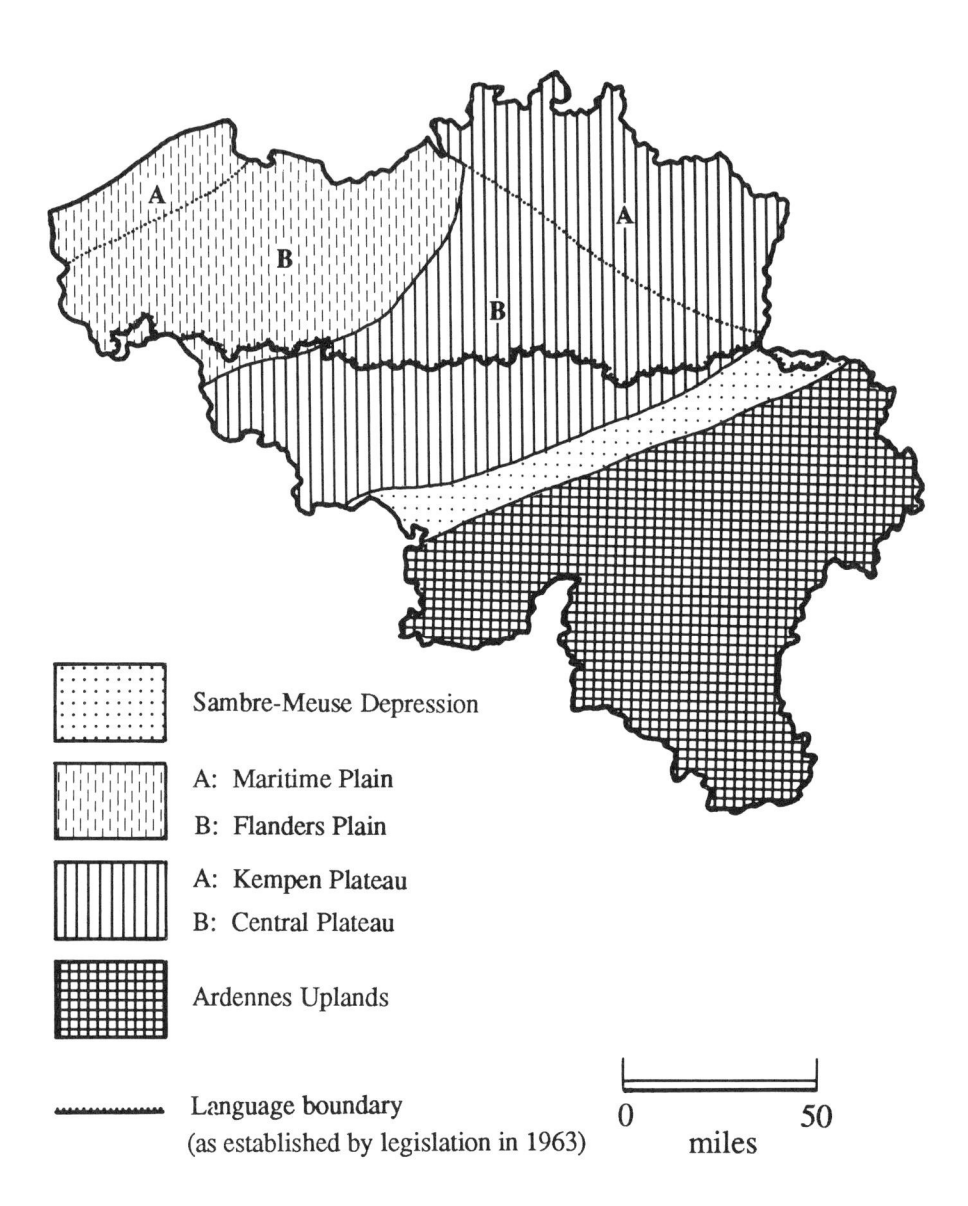

Fig. 3. Simplified relief regions of Belgium.

independence, Belgium was not a country divided ethnically or regionally along language lines. Belgium's subsequent linguistic polarization must therefore be understood as a social and territorial development that largely took place after the founding of the Belgian state.

Chapter 4

AWAKENING ETHNOLINGUISTIC IDENTITY IN THE NINETEENTH CENTURY

The nineteenth century saw the development of a Flemish movement in Belgium and the beginnings of a Walloon reaction to the demands of activist Flemings.[1] In the scholarly literature these events are sometimes treated as reflections or consequences of generalized regional differences between Flanders and Wallonia of a nonlinguistic nature. Mandel, for example, in his efforts to demonstrate the importance of economic factors in the rise of substate nationalism, asserts that during the nineteenth century "Belgium was only half industrialized, for Flanders remained essentially agricultural."[2] That the characteristics of the industrialized portion of the country (encompassing only a small part of southern Belgium) are imparted to the entire Francophone part of Belgium demonstrates the power of subsequently evolved regional divisions to shape thinking about the past. Nonlinguistic regional differences were of undeniable significance in the evolution of the language problem, but sweeping oversimplified, undifferentiated generalizations about the economic and social characteristics of language areas can obscure important questions about the acquisition of conceptual, functional, or political unity by places sharing common linguistic attributes, and about the role of

---

[1] Among the most important histories of the Flemish movement in the nineteenth century are Hendrik J. Elias, *Geschiedenis van de Vlaamse Gedachte, 1780-1914*, 4 vols. (Antwerp: De Nederlandsche Boekhandel, 1963-1965); Paul Fredericq, *Geschiedenis der Vlaamsche Beweging*, 3 vols. (Ghent: Vuylsteke, 1906); Paul Hamelius, *Histoire politique et littéraire du mouvement flamand au 19e siècle* (2nd ed. Brussels: L'Eglantine, 1925); Shepard B. Clough, *A History of the Flemish Movement in Belgium* (New York: Richard R. Smith, 1930).

[2] Ernest Mandel, "The Dialectic of Class and Region in Belgium," *New Left Review*, 20 (Summer 1963), p. 8. This is only one of a host of works crossing the disciplinary and philosophical spectrum that treat Wallonia and Flanders as unproblematic regional entities in the nineteenth century.

territoriality in reflecting and shaping the developing cultural and political geography of the state.

Flanders and Wallonia (as these toponyms are presently understood) were not regional entities as such in the newly created Belgian state. There is little evidence that people living in widely separated parts of northern or southern Belgium felt any regional solidarity with one another, and none of the early Belgian leaders regarded linguistic distribution as a basis for policy distinctions or administrative divisions. Moreover, the plethora of dialects throughout Belgium and the presence of a small but powerful Francophone minority in the North meant that there was no real linguistic unity within the Germanic or Romance areas. Hence, before generalizations on the basis of language distribution are made, we must ask how and why language became the basis for important regional divisions in Belgium as opposed to, for example, internal administrative divisions or economic differentiation.

During the course of the nineteenth century the rise of the Flemish movement, and the reaction of some Walloons to it, set Belgium on the path of linguistic regionalism. Up until 1900, however, the issues, strategies, and concerns were primarily nonterritorial in nature. In fact, by the turn of the twentieth century, the division of Belgium into language regions was only at an embryonic stage of development. By tracing the evolution of regional/territorial conceptions in Belgium, this chapter seeks to bring into question the unproblematic impartment of regional integrity to the Romance and Germanic language areas in the nineteenth century. A chronology of the more important developments during the nineteenth and twentieth centuries appears at the end of the work to assist the reader in following the chain of events.

Characteristics of the Newly Independent Belgian State

The architects of the Belgian state, inspired by the ideals of nineteenth century liberalism, set about establishing Belgium as a parliamentary democracy with a constitutional monarch. The Belgian Constitution, promulgated in 1831, drew heavily on its French counterparts from 1791 and 1830, as well as the American Constitution.[3] Legislative power was vested in a parliament, the king and his elected ministers were entrusted with the executive power, and judicial power was to reside in a court system in which judges were to be named by the king and his ministers. The king's power was limited by a requirement that all royal acts be countersigned by

---

[3] See generally, John Gilissen, "La constitution Belge de 1831: ses sources, son influence," *Res Publica*, 10, special supplement (1968), pp. 107-141; Centre de Recherche et d'Information Socio-Politiques, "Communautés et régions en Belgique: genèse, compétences, institutions, moyens." *Dossier Pédagogique du C.R.I.S.P.* (September 1984).

his ministers. Although the franchise was extended only to large property holders, the Belgian Constitution was hailed in Europe as a progressive document containing significant protections of individual liberties.

As to the choice of a king, internal opposition to the candidacy of the prince of Orange along with objections by England (whose support of Belgian independence had ensured the survival of the new state) to a son of the French king Louis-Philippe led the national congress to offer the position to Prince Leopold of Saxe-Coburg-Gotha.[4] Leopold's mother tongue was German, he was considered an English prince by his first marriage, he had fought against Napoleon in 1813 and 1814, and he then married the daughter of the French king, Louise of Orléans. Consequently, he was a candidate who was acceptable to many possible factions.

In view of the central role played by the Francophone population in the revolt against the Netherlands and the elevated position of French in Belgian society at the time, it is not surprising that the provisional government set up after the revolution used French in its deliberations.[5] The views of the provisional government on language use are evidenced by a series of decrees issued between October and December 1830 making French the language of commands and administration in the military,[6] requiring a knowledge of French (along with history and geography) for candidates seeking positions as officers in the artillery,[7] mandating that the official reporter of Belgian laws be published in French,[8] and eliminating chairs in Dutch language and literature at the Belgian universities.[9] As

---

[4] See generally, Henri Pirenne,*Histoire de Belgique*, vol.7 (Brussels: Lamertin,1932), pp.11-27.

[5] The deliberations were not published in an official journal. The standard reference source on the provisional government, however, is Emile Huyttens, *Discussions du Congrès National de Belgique, 1830-1831*, 5 vols. (Brussels: Société Typographique Belge, 1844-1845).

[6] Decree of October 27, 1830, art. 4, *Bulletin des Arrêtés du Gouvernement Provisoire*, 1, 13 (October 31, 1830), p. 49.

[7] Decree of November 14, 1830, art. 2, *Bulletin des Arrêtés du Gouvernement Provisoire*, 1, 31 (November 18, 1830), p. 121.

[8] Decree of November 16, 1830, art. 1, *Bulletin des Arrêtés du Gouvernement Provisoire*, 1, 33 (November 20, 1830), p. 129. The justification given for this was that the use of Flemish (the term used by the Francophones for the language(s) of the Flemish people) and German varied from province to province, making it impossible to publish official texts in those languages. The decree did provide, however, that (1) in the provinces where other languages were spoken, the provincial authorities would provide translations of laws applicable to all of Belgium and to the province in question (art. 2); (2) citizens could use French, Flemish, or German in dealings with the government (art. 5); and (3) in judicial proceedings a litigant could use his or her mother tongue if the judges and lawyers understood it (art. 6).

Curtis points out, the language policies of the new government represented both a reaction to William's attempt to impose the Dutch language on the southern Low Countries and a continuation of the idea that a single language is necessary for nation building.[10]

The Belgian Constitution provided that "the use of the languages of Belgium is open to choice; it may be regulated only by law, and then only with regard to the actions of public authorities and judicial matters."[11] Although this provision theoretically assured individual freedom of language use, as Polasky has pointed out it effectively promoted frenchification because the language groups were not in an equal position within the state.[12] It in no way protected the already disadvantaged position of the speakers of Dutch and Flemish dialects from the influence of the dominant French language and culture. In addition, the suffrage provisions of the Belgian Constitution primarily benefited Francophones by granting the franchise to only a small percentage of the population based on property holdings.[13] Much of the wealth was concentrated in the industrialized Sambre-Meuse valley and among the upper classes in the cities, which, even in the North, were primarily Francophone. As a result, the political power within the newly created Belgian state was vested almost entirely with French speakers.

The concentration of power in the hands of a largely Francophone upper class did not translate into a state divided into language regions, however. It has already been pointed out that the Belgian state was not bifurcated into two homogeneous linguistic zones. Rather, some of the strongest supporters of frenchification came from the social elite in the North,[14] and dialectical differences hindered communication within the

---

[9] Decree of December 16, 1830, *Bulletin Officiel des Décrets du Congrès National de la Belgique et du Pouvoir Exécutif*, 1, 63 (December 20, 1830), no. 542. Article 17 also provided that examinations were to be held in French unless the student preferred Latin.

[10] Arthur E. Curtis, "New Perspectives on the History of the Language Problem in Belgium" (Ph.D. dissertation, University of Oregon, 1971), p. 192.

[11] Belgian Constitution of 1831, art. 23, *Pasinomie*, 3rd. ser, I (1831), p. 185 ("L'emploi des langues est facultatif; il ne peut être réglé que par la loi et seulement pour les actes de l'autorité publique et pour des affaires judiciaires"). (Author's translation).

[12] Janet Polasky, "Liberalism and Biculturalism," in *Conflict and Coexistence in Belgium: The Dynamics of a Culturally Divided Society*, ed. Arend Lijphart (Berkeley: Institute of International Studies, 1981), pp. 34-45.

[13] The suffrage requirements were so restrictive that in Turnhout, for example, only 699 of the 86,564 inhabitants had the franchise in 1833. A. Van de Perre, *The Language Question in Belgium* (London: Grant Richards, 1919), p. 163. For a general discussion of the suffrage laws see Kenneth D. McRae, *Conflict and Compromise in Multilingual Societies: Belgium* (Waterloo, Ontario: Wilfrid Laurier University Press, 1986), pp. 174-175.

Germanic and Romance linguistic zones. In addition, the early leaders of Belgium did not consider the linguistic geography of the state a basis for territorial divisions. Thus, when the provisional government set about creating military districts for Belgium, it did not hesitate to put the province of Limburg and the province of Liège in the same district.[15] Finally, the highly centralized and unitary character of the Belgian state in its early years militated against the establishment of strong regional divisions. On a formal level, the primary administrative subdivisions of Belgium were the provinces, which did not correspond to language zones. Since the administrative structure of the country was specified in the Belgian Constitution, alternative arrangements could not be brought about without engaging in the difficult and complex process of constitutional revision as provided in article 131 of the Constitution.

The major political divisions in the early years of the Belgian state arose out of differences in religious and political philosophy rather than in language.[16] These divisions, which were to overshadow all other political differences throughout the nineteenth century, were represented institutionally by a Catholic and a Liberal party, with a Socialist party forming somewhat later. Economically, the Belgian textile industry was concentrated in the predominantly Flemish provinces of East and West Flanders, while heavy industry was found largely in the Sambre-Meuse district in the south (figure 4). Although there is obvious linguistic significance to this distribution, neither economic development characterized more than a part of the territories of the two major language families.

In sum, the newly created Belgian state was not divided territorially from a conceptual, functional, or political standpoint into language regions. Despite a constitutional guarantee of freedom of linguistic use after the

---

[14] The Francophones in Flanders could be more vehement proponents of frenchification than the French-speaking Walloons. Thus, Becquet alludes to the following comment of the Flemish Count Robiano de Bornsbeek in 1840: "For Belgians, national unity and intellectual progress in domains such as literature and the sciences are inextricably tied to the spread of knowledge of French." Charles Becquet, "Interaction des ethnies dans la Belgique contemporaine," *Journal de la Société de Statistique de Paris*, 104, 4-6 (April-June 1963), pp. 104-106 (Author's translation). ("Pour nous Belges, l'unité nationale, l'union nationale, le progrès intellectuel, les sciences, la littérature . . . tout cela est lié intimement à l'extension et à la connaissance du français.")

[15] Decree of December 28, 1830, art. 2, *Bulletin Officiel des Décrets du Congrès National de la Belgique et du Pouvoir Exécutif*, 1, 74 (December 31, 1830), no. 631.

[16] Val Lorwin argues that the importance of divisions between liberals, Catholics, and, somewhat later, socialists, significantly delayed the emergence of Flemish political claims since these divisions crossed language lines. Val R. Lorwin, "Linguistic Pluralism and Political Tension in Modern Belgium," *Canadian Review of History*, 5, 1 (March 1970), pp. 8-9.

Legend

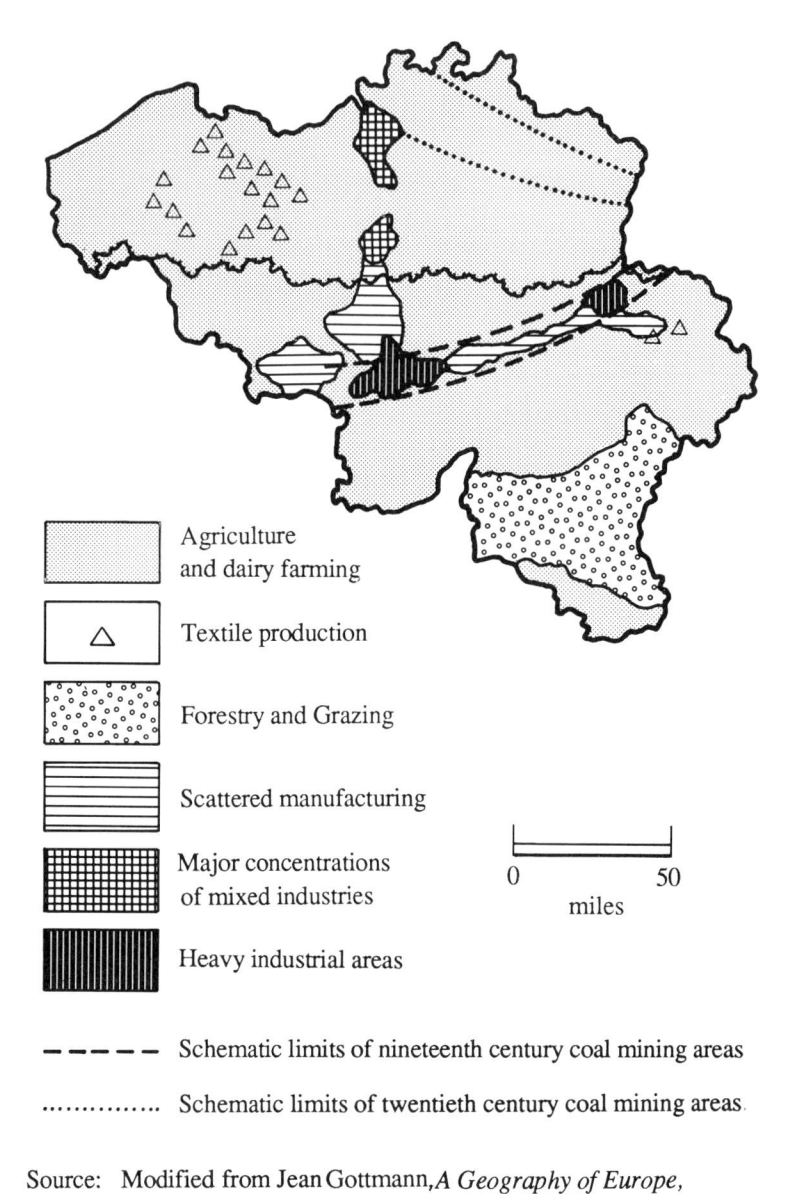

- Agriculture and dairy farming
- △ Textile production
- Forestry and Grazing
- Scattered manufacturing
- Major concentrations of mixed industries
- Heavy industrial areas
- ― ― ― ― Schematic limits of nineteenth century coal mining areas
- .... Schematic limits of twentieth century coal mining areas

0 — 50 miles

Source: Modified from Jean Gottmann, *A Geography of Europe*, 3rd ed. (New York: Holt, Rinehart and Winston, 1962).

Fig. 4. *Simplified economic regions of Belgium.*

Belgian Revolution there was anything but linguistic parity in the country. French speakers dominated positions of political and economic power, and French was the language of administration, the army, the civic guard, the courts, and, perhaps most importantly, the secondary schools and universities.[17] It is not surprising that within a few years some of the Flemish intelligentsia had begun to react to the disadvantaged position of the Dutch language and its speakers in the Belgian state.

The Rise of the Flemish Movement

The Flemish movement originated in the 1830s among a small group of Flemish intellectuals who were swept up by the romantic revival in Europe and concerned about the growing dominance of French in Belgium. The early leaders were middle-class intellectuals, primarily from Ghent and Antwerp, who felt a love for their mother tongue and a newly awakened pride in their cultural heritage.[18] In seeking to popularize their cause, they promoted such symbols of Flemish culture as the lion from the coat of arms of the Counts of Flanders and the story of the Flemish defeat of the French at the Battle of the Golden Spurs in 1302.[19] In fact, many date the formal beginning of the Flemish movement to the publication in 1839 of *The Lion of Flanders* by the Antwerp romanticist Hendrik Conscience describing the battle between the Flemish and the French. Symbols such as these had historical significance only for the inhabitants of the western part of present-day Flanders, and therefore were not necessarily meaningful to those living in northeastern Belgium. Nevertheless, the leaders of the Flemish movement sought to extend their significance to all Belgians speaking Dutch or Flemish dialects.

The Flemish movement at the time was decidedly not separatist. As Ruys notes: "None of these Flemish leaders dreamed of an independent Flanders. They accepted the reality of Belgium and merely asked of the new State that it should not be entirely French-speaking and impose French language and French culture on all others."[20] This orientation is reflected

---

[17] In the North instruction in the primary schools continued to be conducted in Dutch, however. Albert Dauzat, "Le déplacement des frontières linguistiques du français de 1806 à nos jours," *La Nature*, 2775 (December 15, 1927), p. 532, fn. 1.

[18] The emphasis of the movement on literary and cultural issues is suggested by the title of an early pamphlet by one of its early leaders, Ph. Blommaert, "Remarks on the Neglect of the Netherlandic Language" ("Aenmerkingen over de Verwaarlozing der Nederduytsche Tael"), cited in Clough, *A History of the Flemish Movement*, p. 57.

[19] See generally, Clough, *A History of the Flemish Movement*, pp. 65-74.

[20] Manu Ruys, *The Flemings: A People on the Move, a Nation in Being*, trans. from the Dutch by Henri Schoup (Tielt: Lannoo, 1981), p. 45.

in the title and contribution policy of a review founded by a group of Flemish intellectuals which was devoted to Flemish archeology, art history, and medieval literature. It was entitled *Belgisch Museum*, and it accepted contributions from all over the country.[21]

Although the early Flemish leaders were, for the most part, moderate in their demands, they were influential in the development of a Flemish ethnic consciousness. In the preface to *The Lion of Flanders*, Conscience sought to rally Flemings to the Flemish cause by arguing that "there are twice as many Flemings as there are Walloons. We pay twice as much in taxes as they do. And they want to make Walloons out of us, to sacrifice us, our old race, our language, our splendid history, and all that we have inherited from our forefathers. No, there is too much true Flemish blood in the world to allow that, regardless of the measures which the Walloons take."[22] By framing the problem in this way, Conscience was clearly trying to forge a sense of ethnic consciousness among the Flemish peoples in order to promote the cause of the language and culture indigenous to northern Belgium. Focusing on the role of the Walloons, rather than on that of the Flemings who had adopted French ways, put the early leaders in a better position to promote the cause of national consciousness among Flemish people.

The nascent Flemish movement had little initial impact on Belgian policy. In 1835 the king decreed that lectures in institutions of higher learning were to be in French,[23] and French influence continued to grow. In response, the leaders of the Flemish movement began formulating a more specific program of objectives. The first major political initiative was a petition that was drawn up and submitted to the Belgian parliament in 1840. The petition demanded the use of Dutch in the conduct of official affairs in the Flemish provinces and in correspondence between the central government and the Flemish provinces, the establishment of a Flemish academy, and the elevation of Dutch to a position equal to that of French at the University of Ghent.[24] Although the petition evoked little response, it represents the beginning of an articulated and publicized concern for "the Flemish provinces" as a territorial unit within the Belgian state with distinctive needs resulting from linguistic inequalities. The leaders of the

---

[21] Hamélius, *Histoire politique et littéraire du mouvement flamand* , p. 108.

[22] Translation from Clough, *A History of the Flemish Movement*, p. 71.

[23] Royal Decree of December 3, 1835, *Bulletin Officiel des Lois et Arrêtés de la Belgique*, 5, 338 (December 4, 1835), np. During the hearings on the bill, the minister of the interior, Rogier, took the position that a unilingual university was necessary to nation building.

[24] The petition is reproduced in Paul Fredericq, *Geschiedenis der Vlaamsche Beweging*, vol. 1 (Ghent: J. Vuylsteke, 1906), pp. 22-23.

Flemish movement made no effort for some time to define more precisely the meaning of "the Flemish provinces," an inherently ambiguous term given the inexact correspondence between provincial boundaries and language zones. This is not surprising, however, in view of the movement's focus on individual language rights as opposed to territorial issues.

As the small but increasingly visible Flemish movement continued to press its linguistic and cultural claims, its concern with the different character of the Flemish provinces gained limited wider acceptance, particularly among Flemish intellectuals in cites such as Antwerp and Ghent who were sympathetic to the aims of the movement. Ideas about the regional duality of Belgium based on ethnolinguistic criteria even began to gain some ground in official circles, as reflected in a royal decree of 1843 calling for the establishment of two normal schools for primary education in Belgium, one in "the Flemish provinces" at Lier and one in "the Walloon provinces" at Nivelles.[25] Two years later the central government agreed to provide translations of certain laws and decrees for the Flemish communes.[26] These developments represent little more than an acknowledgment of the linguistic duality of the country, however. The Flemish movement itself was still in its infancy and did not enjoy support throughout the North. In fact, a toponym had not yet even become widely adopted for the Flemish part of Belgium.[27]

Although the census of 1846 indicated that the speakers of Dutch or Flemish dialects were in the majority in Belgium, by the middle of the nineteenth century the Flemish movement had not wrested any significant political power from the hands of the Francophones. The economic crisis of the late 1840s had occupied the center stage for a number of years. As it hit the provinces of East and West Flanders the hardest, a few of the leaders of the Flemish movement began equating economic disparities with the Flemish problem. The Flemish movement remained fundamentally concerned with language issues, however. Perceived North-South economic dichotomies grew primarily out of linguistic patterns rather than the reverse.

---

[25] Royal Decree of April 10, 1843, *Bulletin Officiel des Lois et Arrêtés de la Belgique*, 13, 101 (April 11, 1843), np.

[26] Law of February 28, 1845, *Moniteur Belge*, 15, 59 (February 28, 1845), p. 473.

[27] When Minister of the Interior Rogier spoke to the House of Representatives in 1847 about "the Flanders question," he was referring solely to the economic problems of the provinces of East and West Flanders. *Annales Parlementaires—Chambre—1847-1848*, Séance du 4 décembre 1847, pp. 200-201.

Although governmental efforts to cope with the economic crisis of the late 1840s had few directly linguistic overtones, the minister of the interior called for the study of French in Flanders so as to facilitate the migration of Flemish workers to the more prosperous parts of the South.[28] This action reflected the persistence in Francophone circles of the idea that spreading the knowledge of French was necessary to the creation of a unified state. A representative to Parliament stated in 1849:

> As long as Belgian young people are not educated along the same lines, as long as the two races . . . have not, by sharing a common education, effected an intellectual fusion, we will always have two races, and we will never have a nation, possessing one common character, one common spirit, one common name; we will have . . . Flemings and Walloons, but we will not have Belgians.[29]

By the middle of the 1850s, the Flemish activists' message of Fleming-Walloon duality was gaining ground in the North, giving some the confidence to become more radical in their demands. In response to the situation, the government in 1856 appointed a commission to investigate "the most appropriate measures to assure the development of Flemish literature and to regulate the use of Flemish (Dutch) in relations with various parts of the public administration."[30] The government took the surprising step of appointing several noted Flamingants (a term used by Francophones to refer to the more radical Flemish activists) to the commission. When the report was presented, the Francophone-controlled Belgian government regarded it as dangerously radical.[31] The government refused to publish the report at first, and members of Parliament expressed concern that it would lead to discord. Not surprisingly, the report was to become the manifesto of the Flemish movement in the nineteenth century.

The most important comments and recommendations of the committee report concern education, the government, the military, and the diplomatic service. The report called for equality of language in primary

---

[28] *Ibid.*, pp. 200-204. In fact, during this period many Flemings did migrate to the south and, learned French out of necessity.

[29] Remarks of A. Orts, *Annales Parlementaires—Chambre—1848-1849*, Séance du 19 juin 1849, p. 1617. (Author's translation).

[30] Royal Decree of June 27, 1856, *Moniteur Belge*, 26, 192 (July 10, 1856), p. 2509. (Author's translation). The commission was appointed at this time in part to hold off any possible disturbances in connection with the celebration of the 25th anniversary of the ascension of Léopold to the Belgian throne.

[31] *Commission flamande: Installation, délibérations, rapport, documents officiels,* publiée sous la surveillance des membres de la commission (Brussels: Korn, Verbruggen, 1859).

education in Brussels, noting that the grade schools had become heavily frenchified since 1830. In secondary schools, it recommended basic instruction in Dutch in the Flemish provinces with freedom of choice with respect to language thereafter, and an equality of position for Dutch in the Walloon schools equal to that of French in the Flemish schools. At the university level, the report took note of the formation of student groups promoting Flemish culture at the universities in Ghent, Leuven, and Brussels. It did not, however, suggest the creation of a Dutch language university, calling instead for obligatory courses for Flemish students in Flemish history and literature, particularly at Ghent. In terms of government and administration, the commission called for (1) linguistic capability in the central government in both languages so that in dealings with the government, citizens could use the language of their choice; (2) an official translation into Dutch of the *Moniteur Belge* (the reporter of Belgian laws and decrees); (3) the use of Dutch, or Dutch and French, in documents and letters involving the Flemish provinces; and (4) the conduct of judicial proceedings in the language of the parties to the litigation. As for the military, the report sought the establishment of separate Walloon and Flemish regiments in the army and the use of Dutch in the navy in view of the overwhelming number of Flemings in that branch of the service. Finally, the report recommended that the Belgian diplomatic service be able to handle both French and Dutch for the benefit of all Belgians seeking assistance outside of the country.[32]

Despite the negative reaction of the Francophone-dominated Belgian government to the report, it can hardly be regarded as a radical document. In the first place, it dealt solely with linguistic issues and maintained the principle of freedom of individual choice. As such, it reveals the essentially nonterritorial nature of the strategies and priorities of the Flemish movement at the time. There is no suggestion of any kind of a separate political or administrative status for the Flemish provinces, and the term Flanders never appears. Moreover, the central concern with equality of the Dutch and French languages in Belgium is promoted primarily through recommendations for increased bilingualism rather than spatially segregated unilingualism, with the single exception of the proposal for separate Walloon and Flemish regiments in the army. Finally, the commission did not even insist on the exclusive use of Dutch in the Flemish provinces, accepting the presence of Francophone universities in the North and bilingual official documents and correspondence relating to the Flemish provinces. Nevertheless, the Belgian government saw the report as a threat to the unity of the country and promptly issued a

---

[32] See *Commission flamande*, pp. 105-130.

counterreport refuting many of its recommendations.[33] More importantly, it made no effort to follow through on any of the report's suggestions, thereby embittering many of the Flemish activists and pushing the Flemish movement in a more political direction.[34]

The 1850s and 1860s juxtaposed the central government's efforts to build a nation through the integration of peoples against the Flemish movement's attempt to forge a sense of ethnolinguistic identity in northern Belgium. The government pursued its policies of integration on more than a linguistic level. As already noted, it encouraged workers to migrate from poorer to richer regions irrespective of linguistic concerns. Many unemployed Flemish workers moved to the more prosperous sections of the South during this period and eventually adopted the local language and customs. In addition, projects were initiated to improve the economic and social integration of the various parts of Belgium through the construction or improvement of communication and transportation networks. Many of these projects involved the building of railroads to link the North with the South, as revealed by a series of maps of construction projects from the period 1845-1865.[35] The linguistic geography of the country does not appear to have played any role in this development, unless it was to act as a catalyst for projects linking the Germanic and Romance language zones in an effort to discourage the development of functional or conceptual regionalization along linguistic lines.

The priorities of the Flemish movement at the time were largely those contained in the previously discussed commission report. The Flemish activists had been divided ideologically into Liberal and Catholic

---

[33] Curtis, "New Perspectives on the History of the Language Problem in Belgium," p. 221.

[34] In many later Flemish accounts of the movement, the negative reaction to the commission's report is presented as a turning point in the evolution of the Flemish cause. See, for example, M. de Vroede, *The Flemish Movement in Belgium*, trans. from the Dutch by W. Sanders (Antwerp: Kultuurraad voor Vlaanderen & Institut voor Voorlichting, 1975), p. 33.

[35] These maps, in the map collection of the Bibliothèque Royale Albert I in Brussels, include: "Projet de chemin de fer de Jemmapes à Nieuport—carte générale du tracé" (Brussels: Etablissement Géographique de Bruxelles, 1845); "Jonction des trois bassins houillers du Hainaut et des Flandres—chemin de fer de Braine-le-Compte à Gand" (Brussels: Etablissement Géographique de Bruxelles, 1850); "Chemin de fer projété de Tournay à St. Ghislain" (Brussels: Etablissement Géographique de Bruxelles, 1850); "Carte des chemins de fer concédés à la Compagnie Hainaut et Flandres" (Brussels: Etablissement Géographique de Bruxelles, 1858); "Complément de la fusion des intérêts entre les provinces wallonnes et les Flandres—réseau des chemins de fer de l'Ouest Belge ou ligne de Braine-le-Compte à Courtrai avec embranchements sur Quenart et sur Gand" (Brussels: Etablissement Géographique de Bruxelles, 1861); and Erasme Cambier, "Chemin de fer de Tournai à Audenarde" (Brussels: Etablissement Géographique de Bruxelles, 1865). The term *les Flandres*, which appears in some of the titles, referred only to the provinces of East and West Flanders.

camps almost from the beginning. This led them to seek redress for their grievances through the existing political parties rather than through a party of their own. With increasing politicization of the Flemish movement and growing resentment over the response to the commission report, however, Flamingants succeeded in the 1860s in joining forces to win a few parliamentary elections, most notably in Antwerp. Moreover, the Flemish wings of the traditional parties began bringing linguistic issues with regularity before Parliament. In an important symbolic gesture, for the first time a Flemish deputy, Jan de Laet, took the oath of office to the House of Representatives in Dutch.[36]

The leaders of the Flemish movement were increasingly calling attention to "the Flemish provinces" as a part of the Belgian state with special needs and concerns. Although discourse centered around language issues, it sometimes spilled into other areas as well. In one particular oration before the House of Representatives, the Flemish deputy Baron de Maere from Ghent spoke at some length about the disadvantaged economic position of the Flemish provinces. De Maere's characterization of the economic situation was not altogether accurate, but his remarks reveal that language distribution was beginning to be thought of by some as a basis for generalizing about other problems.[37] De Maere's statement was also significant because he pointed to the division in the Flemish provinces between the disadvantaged masses and an aloof Francophone upper class as one of the causes of Flemish economic woes. This was one of the earliest instances in which a Flemish leader directed attention to the social aspect of the Flemish problem rather than to the opposition between Flemings and Walloons.

The impact of the Flemish movement was widening by the early 1870s as the growing Flemish middle class confronted obstacles arising out of linguistic inequalities. Accompanying this development, and directly attributable to the propaganda of the Flemish movement, ideas about the regional distinctiveness of the Flemish part of Belgium gained ground, at least in governmental, political, and intellectual circles. Flemish writers of the period extolled the linguistic and cultural attributes of the Flemish provinces, and Flemish politicians and community leaders confronted the political parties and central government with the special problems of northern Belgium. This does not mean, however, that the language areas of Belgium were well-defined, distinct regions within the Belgian state. To the contrary, there was no political, administrative, or functional integrity to the language areas, and their conceptual significance was extremely limited.

---

[36] Ruys, *The Flemings*, p. 52. A speech was not made in Parliament in Dutch until 1888.

[37] *Annales Parlementaires—Chambre—1868-1869*, Séance du 14 janvier 1869, pp. 243-246.

The Flemish movement had largely developed in the eastern half of northern Belgium among intellectuals, and its influence was little felt outside of intellectual or political circles or in such areas as Limburg. Certainly there was no well developed sense of regional unity among Flemings of the lower classes at the time. In fact, at the Belgian meeting of the International Association of Workers in 1867, a resolution was adopted in favor of Walloon-Fleming solidarity.[38] A few professors at the universities of Brussels and Ghent had proposed the idea of administrative autonomy for the language regions, but their suggestions were largely ignored, even by more radical elements in the Flemish movement.[39] The fundamentally linguistic objectives outlined in the report of the 1856 commission remained the focus of organized Flemish concern.

Legislative Developments in the Late Nineteenth Century

The first significant legislative success of the Flemish movement came in the early 1870s over the right of Flemish defendants to use Dutch in criminal proceedings. This issue was of particular concern to the Flemish movement because of a few blatant instances of injustice in the 1860s when defendants were convicted in proceedings they could not understand.[40] In a proceeding in 1863 held entirely in French, a Flemish poet was fined and sentenced to prison for publishing a poem without the name of the publisher. On appeal, the defendant's lawyer asked that the case be tried in Dutch. In response, the court not only denied the request; it made the sentence more severe. Far worse was a case two years later in which two Flemings with little knowledge of French who were working in the southern part of Belgium were tried and sentenced to death for murder in proceedings held entirely in French. One year after their execution, the real culprits confessed their guilt. A bill was subsequently introduced in the House of Representatives requiring candidates for the judiciary in the Flemish provinces to demonstrate a knowledge of Dutch. Although the measure was defeated, many Belgians, including many Francophones, were outraged by these injustices.

---

[38] Catherine Oukhow, *Documents relatifs à l'histoire de la première internationale en Wallonie*, cahier 47 (Leuven: Centre de l'Histoire de la Première Internationale en Wallonie, 1867), pp. 257-276.

[39] Maurice-Pierre Herremans, "Bref historique des tentatives de réforme du régime unitaire en Belgique," *Courrier Hebdomadaire du C.R.I.S.P.*, 135 (January 1, 1962), p. 4.

[40] The incidents are described in some detail in Clough, *A History of the Flemish Movement*, pp. 100-101; and Van de Perre, *The Language Question in Belgium*, p. 183.

In the early 1870s, a change in the Belgian government produced a more favorable political situation for the Flemish cause. In addition, the defeat of France in the Franco-Prussian war precipitated a decline in French popularity and influence throughout Europe. A representative from Antwerp thought the time was ripe to introduce a bill calling for the use of Dutch in the courts of the Flemish provinces. Further publicized instances of injustice and a flood of petitions to the House prompted the serious consideration of the bill. Although the bill was considerably weakened before adoption, it was passed in 1873.[41]

The law stipulated that in West Flanders, East Flanders, Antwerp, Limburg, and the Leuven district of Brabant, trials of criminal matters were to be in Dutch unless the defendant requested the use of French. Whichever language was chosen, a dossier was to be prepared with a translation of the proceeding in the other language. The law further provided that experts could use the language of their choice and that judges could use French in consultations among themselves. In the Brussels district, the choice of language was "to depend upon the nature of the proceeding," unless the accused spoke only Dutch, in which case it was to be Dutch. The law expressly did not apply to the Belgian appellate courts.

The 1873 law was greeted with mixed reactions by the leaders of the Flemish movement. Although it represented an important step forward for the Flemish cause, it had been substantially weakened in the process of consideration. Nevertheless, it was the first major legislative initiative that acknowledged the special linguistic needs of the northern part of Belgium. This should not be interpreted as a consequence, or as a precipitator, of significant regional divisions along language lines, however. It was a law concerned with language use, and was therefore tailored to apply to that part of the country in which everyone agreed that Dutch or Flemish dialects predominated. By allowing for freedom of language choice in the Flemish provinces and the continued use of French in the appellate courts, it both sidestepped the necessity of delimiting precisely the extent of the Flemish part of Belgium and insured that northern Belgium would not be isolated from the influence of French.

Not satisfied with their partial success on the 1873 law, Flemish activists began pushing for language legislation in the realms of public administration and education as well. It was during this period that the term *Flanders* began to be used frequently in reference to the entire Flemish area of northern Belgium. The cry "*In Vlaanderen, Vlaams*" (in Flanders, Flemish) symbolized the growing concern of the Flemish movement with frenchification in the North. A small victory came on the educational front in 1876 with the passage of a law allowing degrees to be granted in Flemish

---

[41] Law of August 17, 1873, *Moniteur Belge*, 43, 238 (August 26, 1873), pp. 2565-2566.

literature at the University of Ghent.[42] A far more significant achievement was the law of 1878 requiring that in the Flemish provinces and districts other than Brussels notices and communications intended for the public be either in Dutch or in both French and Dutch.[43] Again, however, the law was only a pale reflection of the original bill, which had called for the use of Dutch by all Flemish municipal and provincial authorities.[44]

The first important legislative development in the educational sphere was the adoption in 1883 of a law stipulating that in the Flemish part of the country, courses in the preparatory section of public secondary schools were to be conducted in Dutch.[45] The law further provided that in the regular secondary school program courses in Dutch were to be conducted in that language, while those in French and English were to be held in Dutch until knowledge of the language being taught was such that classes could be carried on in the target language. In addition, Dutch was to be used in at least two other courses beginning in 1888, and the terminology of mathematics and the natural sciences was to be given in both French and Dutch. A significant loophole in the law, however, was a provision allowing classes to be organized for students who could not follow instruction in Dutch.

Many of the Flemish activists regarded the law as woefully inadequate. Indeed, it was only a small step in the direction of providing upper level education in Dutch, and its application proved difficult in view of the small number of teachers qualified to handle courses in Dutch. In order to deal with the latter problem, a normal school was established at the University of Ghent that was later to become important in the controversy surrounding the language of instruction at the university. The impact of the law was necessarily limited because only approximately one-fifth of Belgian children attended public schools. Most went to Catholic schools, which at the secondary level were conducted almost entirely in French. Efforts to change this situation were totally unsuccessful during the nineteenth century.

Despite the limited advances in education, Flemish leaders continued to push ahead on other fronts. During the latter half of the 1880s they

---

[42] Law of May 20, 1876, *Moniteur Belge*, 46, 145 (May 24, 1876), pp. 1489-1493.

[43] Law of May 22, 1878, *Moniteur Belge*, 48, 144 (May 24, 1878), p. 1581. In the Brussels district, the law called for correspondence between the central government and the communes or particular places to be in Dutch if the commune or place requested the use of Dutch or used Dutch itself in its communications.

[44] The text of the original bill can be found in *Annales Parlementaires—Chambre—1875-1876*, April 6, 1876, pp. 760-761.

[45] Law of June 15, 1883, *Moniteur Belge*, 53, 168 (June 17, 1883), p. 2233.

managed to secure the establishment of a royal academy at Ghent to promote the study of Dutch language and literature[46] and a requirement that Dutch be taught in the military schools so that aspiring officers would have a "sufficient knowledge" of the language.[47] More importantly, growing evidence of the ineffectiveness of the 1873 law on language use in the courts precipitated a movement to amend the law. After a long and bitter struggle, an amended version was passed in 1889 that provided for the use of Dutch in the courts of Flemish communes and the use of the language of the accused by the state.[48] This law was significant not only as an advance for the Flemish movement; by calling for the use of Dutch in all Flemish communes, it compelled the government to specify which communes were Flemish. The resulting list, released in 1891, was the first official attempt to delimit the territorial extent of Flemish Belgium.[49]

The 1890s saw other minor legislative gains by the Flemish movement. These included: (1) an 1890 law stipulating that after January 1, 1895, candidates for judicial positions in the Flemish provinces had to demonstrate their ability to comply with the provisions of the 1889 law on language in the courts;[50] (2) a law passed in 1891 obligating the appellate courts in Brussels and Liège to conduct proceedings in Dutch for those cases in which Dutch had been the language in the court of first instance;[51] (3) an

---

[46] Royal Decree of July 8, 1886, *Moniteur Belge*, 56, 191 (July 10, 1886), pp. 2681-2683.

[47] Law of May 6, 1888, *Moniteur Belge*, 58, 131 (May 10, 1888), pp. 1405-1406. The law also stipulated that after January 1, 1892, as many points were to be awarded for knowledge of Dutch as for knowledge of French on officers' examinations (art. 5). By scoring well on other sections of the examination, however, it was still possible to become an officer without demonstrating a knowledge of Dutch.

[48] Law of May 3, 1889, *Moniteur Belge*, 59, 131 (May 11, 1889), pp. 1381-1382.

[49] Royal Decree of May 31, 1891, *Moniteur Belge*, 61, 163 (June 12, 1891), pp. 1637-1643. The list was based on the results of the 1880 census, and included all of the communes that eventually ended up in the Flemish region after the partitioning of the early 1960s with the exception of Spiere-Hellijn, Everbeek, Bever, Walsbets, Wezeren, and Voeren-St.-Pieters. In addition, the list included the commune of Rosoux-Crenwick (currently a part of the commune of Waremme in Liège), and the following communes that are now part of the Brussels bilingual region: Anderlecht, Auderghem/Oudergem, Berchem-Sainte-Agathe/Sint-Agatha-Berchem, Evere, Forest/Vorst, Ganshoren, Jette, Koekelberg, Molenbeek-Saint-Jean/Sint-Jans-Molenbeek, Uccle/Ukkel, Watermael-Boitsfort/Watermaal-Bosvoorde, Woluwe-St. Lambert/Sint-Lambrechts-Woluwe, and Woluwe-St.-Pierre/Sint-Pieters-Woluwe.

[50] Law of April 10, 1890, art. 49, *Moniteur Belge*, 60, 114 (April 24, 1890), pp. 1121-1135. The law also provided that after January 1, 1895, persons could not be named as professors of history, geography, or the Germanic languages in the academies in Flemish towns unless their diploma established that they had successfully passed an examination in Dutch in at least two subjects, that their dissertation was written in Dutch, and that their doctoral lecture was delivered in Dutch.

1897 law extending the provisions of the 1889 law on language in the courts to proceedings before the disciplinary council of the civil guard and those of the 1878 law on language use in public affairs to the administration of the civil guard;[52] and (4) a law passed in 1898 requiring that laws be published in both French and Dutch.[53] The thrust of all of these legislative developments was to ensure that individual speakers of Dutch had the right to use their own language in the schools, the courts, and in dealings with the state.

These legislative successes reflect the growing political power of the Flemings as a group during the late 1800s. The Flemish movement had infiltrated the middle class in northern Belgium, particularly in urban areas. As a result, during the last quarter of the nineteenth century, a number of Flamingants were elected to Parliament, and the Christian (formerly Catholic) party, which drew substantial support from the North, was forced to take up aspects of the Flemish cause. The central issue was complete equality between the Dutch and French languages. In pursuit of that objective, a few Flemish representatives began to use Dutch in the House of Representatives in 1888, and by the turn of the century, a number of the Flemish representatives spoke in Dutch with regularity before Parliament. Other initiatives signalled a gradual enhancement of the status of Dutch. During the last fifteen years of the nineteenth century, the entire reporter of Belgian laws (*Moniteur Belge*) began appearing in both French and Dutch, the government started issuing bilingual bonds, the inscriptions on government buildings began appearing in both languages, and coins and postage stamps were issued with inscriptions in French and Dutch.[54]

The Flemish cause also benefited during the closing decades of the nineteenth century from an economic upturn in parts of Flanders and from changes in the suffrage laws. The emergence of Antwerp as a major world port and a strengthening of the economy in some parts of the North resulted in a small but tangible erosion of the power of the Francophone business elite.[55] A major reform of the voting laws in 1893 to provide for universal male suffrage gave the Flemings more political clout in view of their numerical majority in the state, although the Francophones still held the upper hand as property holders were permitted to cast multiple votes.[56]

---

[51] Law of September 4, 1891, *Moniteur Belge*, 61, 263 (September 20, 1891), pp. 2821-2822.

[52] Law of September 9, 1897, *Moniteur Belge*, 67, 234 (September 11, 1897), pp. 3873-3888.

[53] Law of April 18, 1898, *Moniteur Belge*, 68, 135 (May 15, 1898), pp. 1997-1999.

[54] See generally, Clough, *A History of the Flemish Movement*, pp. 146-147.

[55] See generally, Lode Claes, "Le mouvement flamand entre le politique, l'economique et le culturel," *Res Publica*, 15, 2 (1973), p. 220.

[56] Val R. Lorwin, "Linguistic Pluralism and Political Tension in Modern Belgium," pp. 1-23.

In response to these developments and the gathering momentum of the Flemish movement, a small number of Walloons became concerned with protecting their own interests, thereby giving rise to the beginnings of a Walloon movement.[57]

Walloon Reactions

As early as 1858 a group of intellectuals in Liège got together to form the first Walloon organization, the Société Liègeoise de Littérature Wallonne. Although the founding of the society may have been encouraged indirectly by the activities of the early Flemish movement, its charter makes clear that the organization's sole purpose was to study, support, and propagate literature in the Walloon tongue.[58] That the society was founded in Liège is not surprising. As the capital of a long independent prince-bishopric, Liège had a tradition of autonomous political and cultural development that put it at the heart of Walloon cultural development. The society was important in stimulating interest in Walloon literature and culture,[59] but it did not encourage social or political opposition between Flemings and Walloons. In fact, in 1880 the society sponsored a contest for a literary work celebrating the fiftieth anniversary of Belgium's independence.[60]

A number of other literary and cultural societies sprang up in the Walloon provinces in the 1860s and 1870s and contributed to the development of a limited sense of cultural identity among a small segment of the Walloon population. It was the legislative successes of the Flemish movement in the 1870s and 1880s, however, that signaled the beginning of organized Walloon political opposition to the Flemish cause. The first Walloon political organization sprang up in 1886 in the heavily frenchified Brussels suburb of St. Gilles, and other organizations soon formed in several of the Walloon cities, as well as in Antwerp, Ghent, and Brussels.[61] The organizations in the Flemish cities were composed primarily of Walloon

---

[57] On the Walloon movement in the nineteenth century, see generally, Emile Jennissen, *Le mouvement wallon* (Liège: La Meuse, 1913).

[58] "Statut et règlement," *Bulletin de la Société Liègeoise de Littérature Wallonne*, 1 (1858), pp. 5-10.

[59] Jules Destrée, *Wallons et Flamands: la querelle linguistique en Belgique* (Paris: Plon-Nourrit, 1923), pp. 83-84.

[60] Société Liègeoise de Littérature Wallonne, *Cinquantième anniversaire de l'indépendance nationale* (Liège: H. Vaillant-Carmanne, 1880).

[61] Paul Fredericq, *Vlaamsch België sedert 1830* (Ghent: Vuylsteke, 1906), pp. 197-199.

civil servants living in the North who were seeking to defend their positions and interests against Flemish incursions.

The reactive character of the nascent Walloon movement is seen in the language of the publications of the movement and of the charters of the societies. In the early 1890s, for example, a few of the promoters of the Walloon cause began publishing a periodical under the title *La Défense Wallonne*. In one of the early issues, the founding charter of a Walloon organization, La Ligue Wallonne d'Ixelles, is reprinted. Article 1 states:

> The goals of the Walloon League of Ixelles are: First—to cement relationships among Walloons living in the Brussels area and to allow them to come to each others assistance; second—to fight all exaggerated demands of 'flamingantisme' and to investigate the means to resolve equitably the question of language usage in Belgium.[62]

Congresses bringing together Walloons from all over Belgium were organized in 1890, 1891, 1892, and 1893. Discussions at the meetings were anti-Flamingant in tone, and centered around the position of French in Belgium, ways of safeguarding the interests of Walloon functionaries, and strategies to encourage the diffusion of French throughout the country.[63] Frequent reference was made during the congresses to Wallonia as a distinct cultural region within Belgium. Although the term had first been used in the 1840s and 1850s among a small group of Liège intellectuals,[64] it was during this period that it took on wider significance. Internal divisions led to the abandonment of an annual congress after 1893, but the movement was only beginning. In fact, just before the turn of the century, a few Walloon leaders began publishing the *Revue Wallonne* "to combat the flamingant peril that threatens the country."[65]

Despite the obvious significance of these early Walloon reactions to the Flemish movement, up until around 1900 the Walloon movement was extremely limited in scope. It was the creation primarily of a few intellectuals who sought to encourage Walloon cultural development and of a group of political figures and functionaries who felt threatened by Flemish advances. Certainly no widespread feeling of Walloon identity developed during this period, and Wallonia had no meaning as a regional entity in the minds of most Belgians. In part this may be attributed to the fact that the Walloons as a group were not noticeably disadvantaged within

---

[63] Fernand Schreurs, *Les congrès du Rassemblement Wallon de 1890 à 1959* (Charleroi: Institut Jules Destrée, 1960), pp. 9-14.

[64] Albert Henry, *Esquisse d'une histoire des mots Wallon et Wallonie* (Brussels: La Renaissance du Livre, 1974), p. 9.

[65] Schreurs, *Les Congrès du Rassemblement Wallon*, p. 15. (Author's translation).

the Belgian state. Despite Flemish advances in certain areas and the suffrage changes of 1893, Francophones continued to dominate the social, economic, and political life of Belgium at the turn of the century, and many of those Francophones were Walloons.

An Overview of Linguistic Regionalism in the Nineteenth Century

As the foregoing historical discussion has suggested, the present regions of Flanders and Wallonia began to acquire a limited degree of conceptual significance during the nineteenth century among a segment of the Belgian population. This development was considerably more important in the North than in the South, but in neither area did linguistic regionalism penetrate significantly to the level of the masses, and even its proponents by and large did not argue for any degree of autonomy or separation for Wallonia and Flanders. A few Belgians had raised the idea of linguistic federalism, even in Walloon circles,[66] but most writings and discussions of the time were premised on the assumption of a unitary Belgian state. Even most of the Flamingants espoused loyalty to the Belgian state.[67] At the end of heated debates in Parliament over language issues, someone would invariably proclaim: "Fleming and Walloon are first names, Belgian is our family name."[68]

The orientation of the Flemish movement in the nineteenth century was emphatically nonterritorial. As De Schryver points out: "The first generations of Flamingants did not ask for the abolition of French in the Flemish areas; they asked only for official recognition of their own language and accepted bilingualism in their region without disputing the French unilingualism of the Walloon provinces. In addition, the Flamingants were loyal to the Belgian fatherland on the whole continuously until World War I; they even considered themselves as more authentic Belgians than the Walloons. There were demands neither for separatism nor for federalist

---

[66] A call for federalism by a Walloon is documented in the article "Flamands et Wallons" reprinted from the June 12, 1890, issue of *La Réforme* in W. Houtman, *Vlaamse en Waalse Documenten over Federalisme* (Schepdaal: Het Pennoen, 1963), document 1. For an analysis of Flemish thinking at the time on federalism, see Arthur De Bruyne, *Het Federalisme in Vlaanderen* (Schepdaal: Het Pennoen, 1962), pp. 190-208.

[67] As the twentieth century historian of the Flemish movement and supporter of the Flemish cause, Maurits Van Haegendoren points out: "In the nineteenth century, Flemish action did not go beyond petitions, manifestations against the wrongs inflicted on the Flemish, and student revolts, often accompanied by violence." Maurits Van Haegendoren, *The Flemish Movement in Belgium*, translated from the Dutch (Antwerp: Flemish Cultural Council, 1965), p. 27.

[68] "Flamand, Wallon, sont des prénoms, Belge est notre nom de famille." Clough, *A History of the Flemish Movement*, p. 151.

reorganization of the Belgian state.⁶⁹ The focus on individual rights issues, as opposed to territorial concerns, is suggested by the lack of attention given to the exact delimitation of "the Flemish provinces" throughout most of the nineteenth century. Even more telling was the lack of concern shown for regional linguistic integrity. In Brussels, census data indicate that by the middle of the nineteenth century the use of the French language was limited primarily to those living in the center of the city, whereas by 1900 it had spread significantly to the south.⁷⁰ As Platel points out, however, the Flemish movement did not focus on the Brussels problem during this period, accepting the growing influence of French as a fait accompli.⁷¹

On the Walloon side, the idea of a unified Wallonia was even more ambiguous. In explaining the composition of the Walloon congress of 1890, Schreurs points out that "at the time, the territorial idea of Wallonia was still confused in the minds of most people."⁷² This is not surprising in view of the lack of a well-developed Walloon sense of ethnic or regional identity at the time. Indeed, for most Walloons feelings of attachment and identity were primarily local (for example, at the level of the town or the commune) during the nineteenth century as they had been for generations. The primary exceptions were members of the business or political elite who had come to think of themselves as Belgians. Thus, outside of the relatively small group of intellectuals who were behind the organized Walloon movement, little concern was manifested during this period for linguistic integrity in the South. Admittedly this was not a major issue in view of the high degree of unilingualism in Wallonia. In response to the continued influx of Flemish workers to the Sambre-Meuse district, however, rather than insisting on the sole use of French, socialist groups started providing French translations of technical terms.⁷³ In fact, the conciliation board of

---

⁶⁹ Reginald De Schryver, "The Belgian Revolution and the Emergence of Belgium's Biculturalism," in *Conflict and Coexistence in Belgium: The Dynamics of a Culturally Divided Society*, ed. Lijphart, p. 27.

⁷⁰ For a detailed discussion of the census data on Brussels in the nineteenth century, see Frank Logie, "Ruimtelijke Spreiding van de Nederlandstalige Bevolking in Brussel-Hoodstad," *Taal en Sociale Integratie*, 2 (1981), pp. 88-94.

⁷¹ Marc Platel, "Bruxelles, aussi capitale des Flamands," *Septentrion—Revue de Culture Néerlandaise*, 9, 2 (June, 1980), pp. 77-81, trans. from the Dutch by Willy Devos.

⁷² ("[A] cette époque, la notion territoriale de Wallonie était encore confuse dans la plupart des esprits.") Schreurs, *Les Congrès du Rassemblement Wallon*, p. 9. (Author's translation).

⁷³ This was done by the Socialist Association of Wood Workers. *Revue du Travail 1900* (Brussels: Ministère de l'Industrie et du Travail, 1900), p. 925. A Socialist party had been founded in 1885 which was particularly strong in the South. Although the party's central focus was on economic, as opposed to linguistic, matters, its Flemish and Walloon leaders differed on language issues, just as did the leaders of the other major Belgian political groups.

industrial disputes of the Ministry of Industry and Work denied employers the right to discharge a worker immediately who, despite a request to the contrary, continued to speak Dutch with a fellow worker.[74]

Whatever developments may have occurred during the nineteenth century to draw attention to the linguistic heterogeneity of the country, Belgium was still a politically and functionally unified state at the turn of the century. Not one of the many Belgian *enquêtes* appearing before 1900 was broken down by language region. Moreover, ambiguity remained in the usage of the term Flanders, with the word still being employed occasionally to refer only to the provinces of East and West Flanders.[75] The subject catalog at Belgium's national library does not contain any listings under the headings "Wallonie," "Wallon," or "Wallons" of works appearing before 1900 with a specific reference to the linguistic region of Wallonia in the title. Only a handful of pre-1900 works appear under the headings "Vlaanderen," "Vlaamse," or "Vlamingen" that use the term Flanders in the title in reference to the Dutch-speaking part of Belgium.[76] This is hardly surprising as the language areas had not yet taken on any functional or formal significance. For example, when the Minister of Agriculture, Industry, and Public Works established demonstration areas for administrative purposes in 1890, he did not hesitate to put the southern provinces of Hainaut and Liège, and the northern province of Limburg in the same region.[77]

The Flemish movement in Belgium developed essentially along class lines in the North, but not in the traditional Marxist sense. The leaders of the Flemish movement were a new bourgeoisie seeking to gain power in a state in which the French language was the key to social and political advancement.[78] As the Flemish movement developed, the language areas

---

[74] Ministère de l'Industrie et du Travail, *Revue du Travail* (Brussels: Ministère de l'Industrie et du Travail, 1896), p. 405.

[75] For example, A. Cosny, *Au beau pays de Flandre* (Brussels: Grands Annuaires, 1902), a tourist manual for Belgians that describes only the cities and towns of West Flanders.

[76] These conclusions are based on an analysis of the card catalog at the Bibliothèque Royale Albert 1er in Brussels that was performed in June 1986, and included the equivalents of the terms cited in the text in the other major national language. It should be kept in mind that the holdings of the national library are by no means complete, and mistakes in the catalog abound. Nevertheless, omissions and mistakes are presumably random, so the analysis should provide some indication of the use of these terms before 1900.

[77] Decree of April 10, 1890, *Moniteur Belge*, 60, 108 (April 18, 1890), pp. 1061-1064.

[78] The same idea has been expressed in Georges Goriely, "Frontière linguistique et destin de la Belgique," in *Introduction à l'histoire des doctrines politiques contemporaines: essais sur le nationalisme*, ed. Georges Goriely (Brussels: Presses Universitaires de Bruxelles, 1982-1983), pp. 70-86.

sometimes became the basis for generalizations about economic, educational, and religious matters. This is understandable in view of the importance of these factors in the evolution of the ethnolinguistic movements. Thus, the concentration of industrial growth in the Sambre-Meuse valley was of significance in perpetuating Francophone dominance, and the frenchification of higher levels of education put native Dutch speakers at a continual disadvantage. As noted at the beginning of this chapter, however, to argue that ethnolinguistic tensions in the nineteenth century were the result of North-South differences is to ignore the true complexities of the situation. Averages by language area may show that northern Belgium was less industrialized, less well educated, and more traditionally Catholic than southern Belgium, but there was no distinct Flanders-Wallonia dichotomy with respect to these variables. Variations within the North and the South were often as great as differences between the language areas. Consequently, the significance of these patterns as independent variables in the rise of linguistic regionalism is problematic at best.

An examination of census data from the nineteenth century at the provincial level reveals the problems of assuming a simple correlation between economic and social patterns on the one hand and gross linguistic distributions on the other. To begin with the economic situation, descriptions of nineteenth-century Belgium sometimes assume a clear-cut dichotomy between an agricultural, backwards Flanders and a prosperous, industrialized Wallonia. Although northern Belgium was in general more agricultural than the South, looking at the issue at the provincial level reveals substantial intraregional variations. It is possible from the 1846 census to calculate the percentage of the total work force in each province engaged in agriculture. Table 3 contains the figures, for each province, arranged in descending order.

The figures in table 3 do not reveal any clear Wallonia/Flanders dichotomy. This is confirmed by figure 5, a provincial map of the country categorizing provinces on the basis of whether greater than or less than 50 percent of their work force was engaged in agricultural activities in 1846. In order to facilitate comparison of the resulting patterns with linguistic distribution, the contemporary language border has been superimposed on the map.[79] In addition, data for the province of Brabant, which straddles the linguistic boundary, have not been included. It is evident that the resulting pattern does not correspond strongly with language distribution. Although

---

[79] The stability of the language border, with the exception of the area around Brussels, allows us to use the contemporary border, which was delimited administratively in the early 1960s, as an approximate indicator of language zones in the nineteenth century.

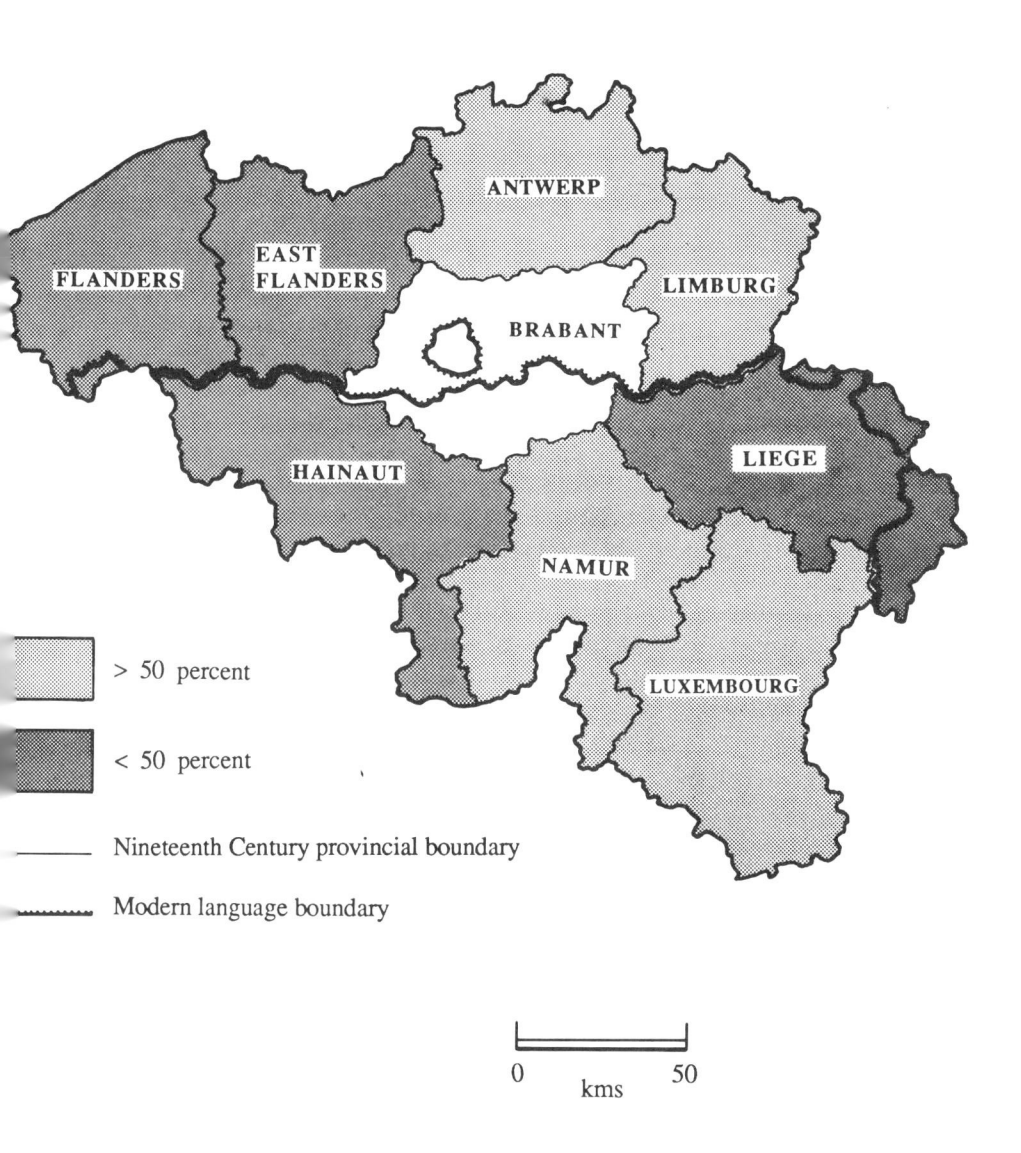

Fig. 5. Percentage of work force engaged in agriculture in Belgium in 1846.

Table 3. Percentage of Work Force in Agriculture, 1846

Province	Percentage of work force in agriculture	Present language region of the province
Limburg	69.4	Flanders
Luxembourg	64.6	Wallonia
Brabant	57.0	Flanders and Wallonia
Namur	56.1	Wallonia
Antwerp	52.9	Flanders
East Flanders	49.8	Flanders
Hainaut	47.2	Wallonia
West Flanders	44.3	Flanders
Liège	43.7	Wallonia

SOURCE: Calculated from: *Annuaire Statistique de la Belgique*, 1870 (Brussels: Ministère de l'Intérieur, 1870), pp. 38-39.

the relative position of some of the provinces with respect to this indicator had changed by the 1856 census, the same map could have been produced based on the results of that census, as revealed by table 4.

The pitfalls of dividing the nineteenth-century Belgian economy along ethnoregional lines are further revealed by data from the 1856 census on the percentage of inhabitants in each province employed in industrial activities. The figures are presented in declining order in table 5. Although the data reveal no close link between language and levels of industrialization, they hide an important economic aspect of nineteenth-century Belgium: the concentration of textiles in the Flemish provinces of East and West Flanders and of heavy industry in the Sambre-Meuse valley.[80] Even taking this into account, however, levels of industrialization in southeastern Wallonia and eastern Flanders were more similar to each other than either was to the language zone to which it belonged.

Turning to educational and religious considerations, the available data do not entirely support the generalization that the inhabitants of northern Belgium in the nineteenth century were less well educated and more devoutly Catholic than their counterparts in the South. Table 6 presents data for the percentage of young Belgians who could not read or write at the time of application for military service in 1867, arranged in ascending order of literacy. Although there is some correspondence between levels of literacy and the language regions, it is hardly complete.

---

[80] See generally, *Recensement Général*, 1856, pp. 228-229.

Table 4. Percentage of Work Force in Agriculture, 1856

Province	Percentage of work force in agriculture	Present language region of the province
Limburg	69.3	Flanders
Luxembourg	68.6	Wallonia
Antwerp	56.0	Flanders
Namur	53.5	Wallonia
Brabant	45.6	Flanders and Wallonia
East Flanders	45.3	Flanders
Hainaut	40.5	Wallonia
West Flanders	39.7	Flanders
Liège	36.3	Wallonia

SOURCE: Calculated from: *Recensement Général (31 décembre 1856)—Population* (Brussels: Ministère de l'Intérieur, 1861), pp. 228-229.

Table 5. Percentage of the Population in Industry

Province	Percentage of population in industry	Present language region of the province
West Flanders	35.5	Flanders
East Flanders	28.1	Flanders
Hainaut	25.3	Wallonia
Liège	23.4	Wallonia
Brabant	18.8	Flanders and Wallonia
Antwerp	13.5	Flanders
Namur	13.0	Wallonia
Limburg	10.0	Flanders
Luxembourg	7.8	Wallonia

SOURCE: Calculated from: *Annuaire Statistique, 1870*, pp. 206-207; 24-27.

Figure 6 depicts cartographically a logical grouping of the provinces into three categories of literacy based on these data, those with illiteracy rates over 30 percent, those with rates between 20 percent and 26 percent, and those with rates under 15 percent. The presence of Hainaut in the category of highest illiteracy casts some doubt on a simple correlation between educational and linguistic patterns.

Table 6. *Percentage of Illiterate Applicants for Military Service, 1867*

Province	Percentage of illiterate applicants	Present language region of the province
Hainaut	35.1	Wallonia
East Flanders	32.8	Flanders
West Flanders	26.0	Flanders
Brabant	24.4	Flanders and Wallonia
Antwerp	24.1	Flanders
Liège	23.1	Wallonia
Limburg	22.1	Flanders
Namur	12.8	Wallonia
Luxembourg	6.4	Wallonia

SOURCE: Calculated from: *Annuaire Statistique, 1870*, pp. 118-119.

With regard to the religious issue, it is difficult to measure levels of adherence to traditional Catholic practices in the nineteenth century because data are not available for indicators such as attendance of mass. Divorce rates can provide some useful information, although the practice was sufficiently uncommon that differences between the provinces are not always statistically significant. Nevertheless, data from 1890 on marriages in which one of the partners is divorced are somewhat revealing. Table 7 sets forth the data in declining order.

Due to the uncertain significance of these figures, it is useful to look at another possible index of religious belief for which statistics are available: the number of inhabitants per Catholic priest in each province. Again, this is not a perfect measure of adherence to Catholicism because the number of priests depends not just on attendance of services, but on such factors as population distribution patterns. Nevertheless, looking at this variable together with the percentage of marriages involving divorcees provides some indication of patterns of adherence to traditional religious practices. The figures by province for 1890 for the number of people per Catholic priest, presented in declining order, are set forth in table 8.

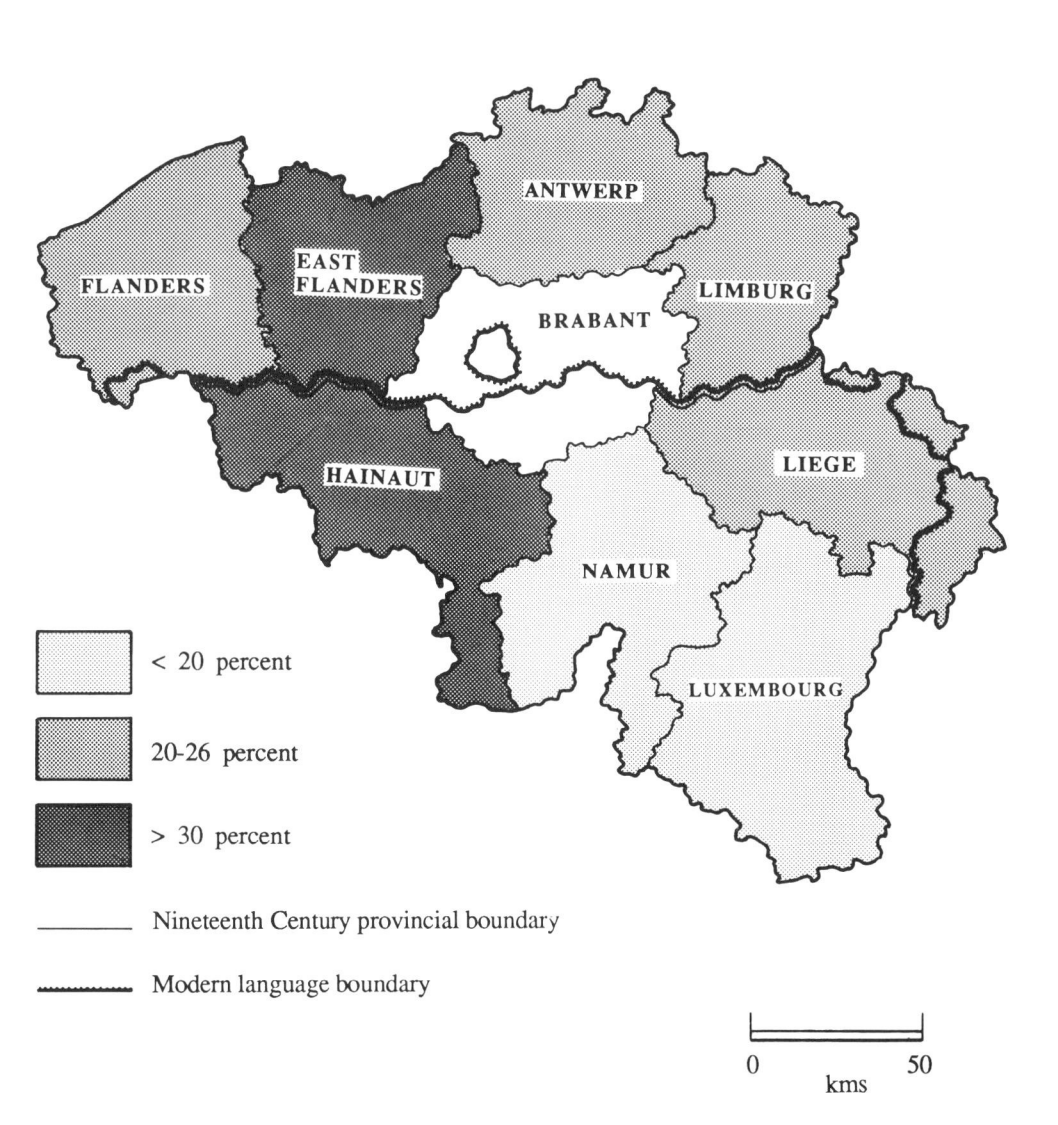

Fig. 6. Percentage of illiterate applicants for military service in Belgium in 1890.

Table 7. Percentage of Marriages Involving Divorcees

Province	Percentage of marriages involving divorcees	Present language region of the province
Brabant	0.84	Flanders and Wallonia
Liège	0.41	Wallonia
Antwerp	0.31	Flanders
Hainaut	0.25	Wallonia
Namur	0.16	Wallonia
East Flanders	0.14	Flanders
West Flanders	0.10	Flanders
Luxembourg	0.07	Wallonia
Limburg	0.00	Flanders

SOURCE: Calculated from: *Statistiques du Mouvement de la Population de l'Etat Civil en 1890* (Brussels: Ministère de l'Intérieur et de l'Instruction Publique, 1895), pp. 666-725.

Table 8. Number of Inhabitants per Priest, 1890

Province	Number of inhabitants per priest in 1890	Present language region of the province
Hainaut	1165	Wallonia
Liège	1066	Wallonia
Brabant	1027	Flanders and Wallonia
East Flanders	934	Flanders
West Flanders	880	Flanders
Antwerp	867	Flanders
Namur	658	Wallonia
Limburg	585	Flanders
Luxembourg	568	Wallonia

SOURCE: Calculated from: *Statistiques de la Population en 1890*, pp. 454-455; 336-337.

Figure 7 is a map that groups the Belgian provinces on the basis of the degree of Catholic religiosity of their inhabitants as reflected by a combined consideration of the data from the two previous tables. In 1890, Limburg, Luxembourg, and Namur had fewer than 800 inhabitants per priest and a divorce rate among marriage partners of less than 0.2 percent. These provinces would presumably be the most traditionally Catholic. Brabant, Hainaut, and Liège, by contrast, had more than 1000 inhabitants per priest and a divorce rate among marriage partners of more than 0.2 percent. These provinces could be considered as the least Catholic in Belgium. All of the other provinces have between 800 and 1000 inhabitants per priest and divorce rates among marriage partners between 0.10 percent and 0.31 percent. These are grouped together in a middle category. An examination of figure 7 reveals once again that social patterns did not necessarily correspond to language distribution.

The data that have been examined present nothing more than a static, simplified picture of economic and social patterns in Belgium at particular points during the nineteenth century. Figures from other censuses or analyses of alternative factors would undoubtedly yield somewhat different patterns. What is evident, however, is that there was no clear or simple correlation between areas sharing economic and social characteristics and the distribution of languages. Thus, during the nineteenth century the province of Limburg had little in common with much of the rest of northern Belgium aside from language in the general sense of the term. Since language was not a widespread basis of significant regional identity in southern Belgium at the time, it makes little sense to treat Limburg conceptually in the same analytical category as East and West Flanders except from a philological point of view, or possibly in reference to the cultural or political objectives of the Flemish movement.

The point is that the growth, timing, and significance of ethno-regional consciousness in the nineteenth century cannot be explained solely by reference to large-scale economic, educational, or religious differences. Differentiation along these lines did not correspond entirely to language patterns and did not always translate into regional consciousness. Rather, these differences provided the context in which ethnic and regional identity was to develop. The Flemish movement in its early years was fundamentally a struggle for linguistic rights. Although linguistic distribution became the basis for limited generalizations about other aspects of Belgium's social, economic, and political character over the course of time,

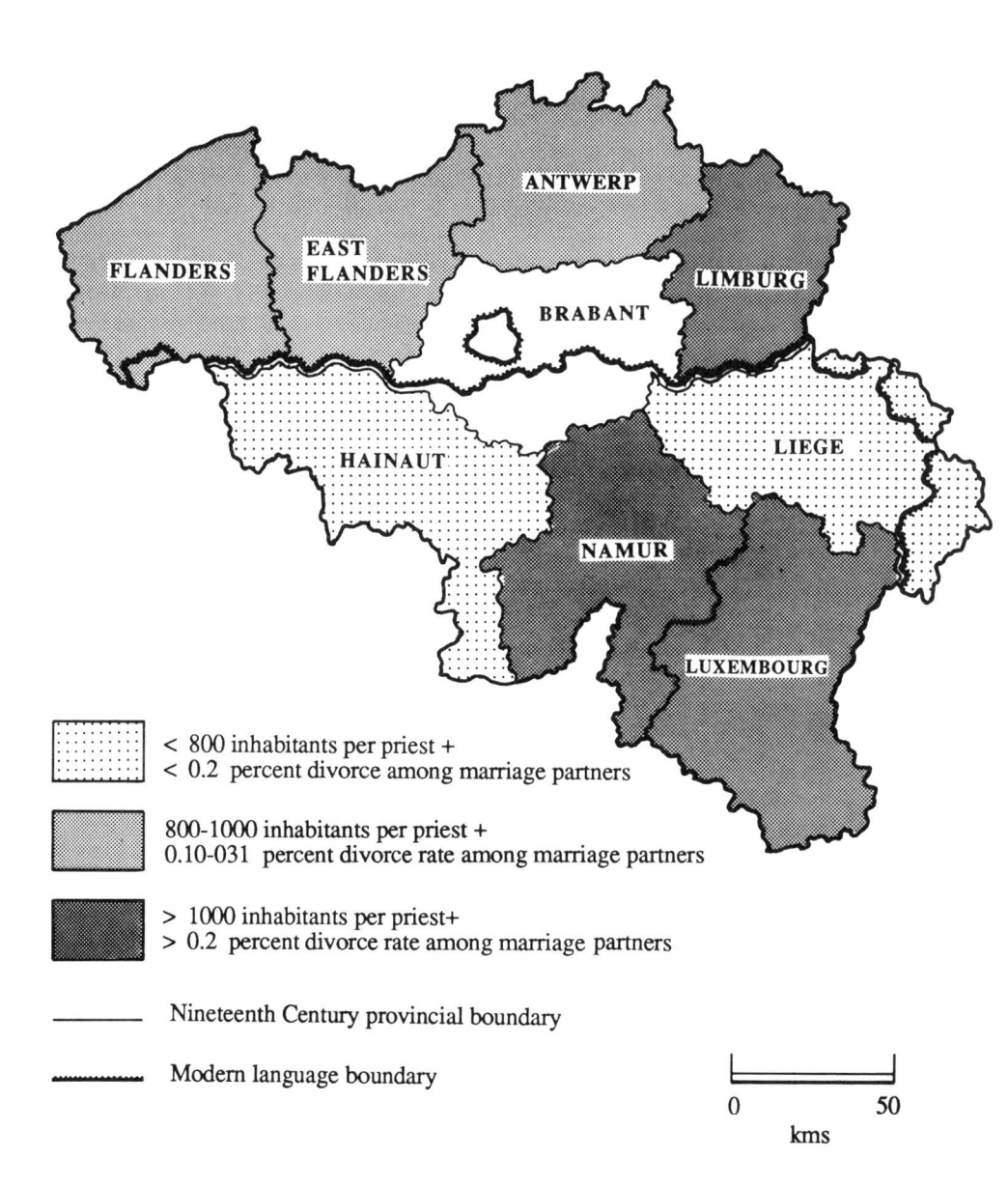

Fig. 7. Indicators of strength of traditional Catholic values in Belgium.

the focus of the Flemish movement and its Walloon counterpart remained primarily linguistic. By the turn of the twentieth century, Wallonia and Flanders had only begun to take on a degree of conceptual significance for a limited sector of the Belgian population. They certainly did not constitute meaningful economic, social, or ethnic territorial units and, consequently, should not be treated as important regional entities in our attempts to understand the historical development of nineteenth-century Belgium.

Chapter 5

THE RISE OF ETHNOREGIONALISM IN THE EARLY TWENTIETH CENTURY

The legislative and organizational successes of the Flemish movement toward the end of the nineteenth century precipitated growing awareness among some Belgians of the ethnoregional duality of their country. The fact that most Dutch speakers continued to be at a social and political disadvantage within the state had not yet translated, however, into a sense of ethnoregional identity in the North. Dialectical differences remained a barrier; there was little historical, political, or functional basis for linguistic regionalism; and the impact of the Flemish movement was still limited. Among the Walloons, ethnoregional identity was even less well-developed, being limited to a few intellectuals and political figures who were concerned that Flemish legislative gains might threaten the position of French and its speakers in Belgium's political and social life.

By the middle of the twentieth century, the language areas of Belgium had evolved into well-defined and widely accepted regional entities, setting the stage for the sweeping partitioning schemes of the past twenty-five years. This development is tied to a fundamental shift in Belgium's ethnolinguistic politics from a focus on individual language rights to an emphasis on the internal integrity of linguistic territories. Although this change is frequently acknowledged in the literature, it is rarely probed in any depth, a reflection of the previously discussed tendency to treat regions as simple reflections of social patterns. Yet the rise of linguistic-territorial issues and strategies during the first half of the twentieth century served to recast the context and nature of group interaction. As such, the evolution of linguistic regionalism deserves explicit analysis.

Language, Ethnicity, and Regionalism prior to World War I

During the decade and a half preceding World War I the leaders of the Flemish movement continued to press for the rights of individual Dutch speakers to use their native language in the schools, the military, the courts, and the government. Although the position of Dutch in Belgium had improved somewhat as a result of the nineteenth century language laws, higher education was still conducted primarily in French, the army was dominated by Francophone officers, and the use of Dutch was rare in the upper echelons of the central government. The issue of education was of particular importance, because advanced training in Dutch was critical to changing the established linguistic hierarchy. This spurred many Flemish leaders at the end of the nineteenth century to initiate a campaign to make the University of Ghent into a Dutch-speaking institution.

The effort was weakened at first by differences in opinion among Flemish leaders as to the extent to which Dutch should be introduced into the university. The campaign gained considerable momentum in 1906, though, with the publication of an influential treatise by Lodewijk De Raet calling for Dutch to be the sole language of the state university at Ghent.[1] In justifying his position, De Raet wrote about the disparate impact of the Industrial Revolution on the Flemish and Walloon parts of Belgium, thereby contributing to the growing conceptual significance of the language regions.[2] The primary issue, however, was the linguistic status of the university itself. The Flemish movement largely united behind De Raet's proposals, and De Raet was appointed as the head of the university commission to investigate the situation. The commission's report, which called for a gradual switch at Ghent to the use of Dutch as the language of instruction,[3] received considerable support from many Flemings. It met with staunch opposition from some of the Francophones in Parliament, however. At the same time, counterproposals to make Ghent a bilingual university were rejected by those Flemish leaders who were adamant in

---

[1] Lodewijk De Raet, *Over Vlaamse Volkskracht: De Vervlaamsching der Hoogeschool van Gent* (Brussels: De Vlaamsche Boekhandel, 1906).

[2] See also Lodewijk De Raet, *Een Economisch Programma voor de Vlaamsche Beweging* (Brussels: De Vlaamsche Boekhandel, 1906). The significance of his ideas will be discussed in more detail below.

[3] Vlaamsche Hoogeschool Commissie, *Verslag over de Vervlaamsching der Hoogeschool van Gent* (Ghent: Algemeen Nederlandsch Verbond, 1909). The report called for all instruction to be in Flemish with the exception of foreign languages, with an eleven-year transition period during which the use of French would be phased out.

their objective of establishing a Dutch-language university. The resulting deadlock had still not been resolved at the outbreak of World War I.

Although De Raet used regional differences to justify his position on the University of Ghent, the thrust of the movement for a Dutch-language university was not primarily territorial. It was an extension of the Flemish movement's concern with providing equal opportunity for Dutch speakers in Belgium. The other major university in Flanders, the Catholic University of Louvain, was entirely French-speaking, with the episcopate in 1906 specifically refusing to offer courses in Dutch.[4] Hence, Ghent was seen as the best hope for educating Flemings in Dutch, and thus preparing them to enter responsible positions in education, business, and government.

The debate over extending the application of the 1883 education law to private secondary schools had greater territorial implications because the original law was based on linguistic distinctions between the Flemish and Walloon parts of Belgium. Many of the Flamingants pushed for the use of Dutch in all Flemish schools, but insufficient support existed in Parliament for such an initiative. A compromise was finally reached in 1910 with a law requiring that university applicants coming from secondary schools in the Flemish part of the country be examined in Dutch unless they had taken at least two nonlanguage courses that had been conducted in Dutch or had spent at least eight hours per week learning Dutch.[5] Although the impact of the law was limited, it was another legislative initiative embodying a distinction between the North and the South. In fact, some Walloons contended that Flemish supporters of the measure were trying to divide Belgium.[6]

In 1914 Parliament also passed a bill requiring attendance at school until the age of twelve.[7] In discussions of the bill, a few Flemings pressed for the designation of language territories in Belgium, with the regional language determining the school language.[8] This idea met with considerable opposition, however, and the final compromise stipulated that

---

[4] Pierre Maroy, "L'évolution de la législation linguistique belge," *Revue du Droit Public et de la Science Politique*, 82, 3 (May-June 1966), p. 460. The episcopate did, however, agree to the introduction of a few courses in Dutch in 1912.

[5] Law of May 12, 1910, arts. 2 & 5, *Moniteur Belge*, 80, 135 (May 15, 1910), pp. 2864-2866.

[6] See Shepard B. Clough, *A History of the Flemish Movement in Belgium: A Study in Nationalism* (New York: Richard R. Smith, 1930), p. 154.

[7] Law of May 12, 1910.

[8] Discussed in Arthur C. Curtis, "New Perspectives on the History of the Language Problem in Belgium," (Ph.D. dissertation, University of Oregon, 1971), p. 293.

education be given in the mother tongue of the child.[9] Although the enforcement of the law posed some problems, particularly in Brussels, it was significant for the Flemish movement both because it discouraged Dutch-speaking parents from sending their children to French-language schools and because it raised the level of literacy of the Flemish population as a whole. The law did not, however, prohibit instruction in French as a second language in the primary schools, a matter of continued concern for many of the more radical Flemings.

With the growing militarism in Europe in the years leading up to World War I, language use in the Belgian army once again surfaced as a prominent concern. The 1888 law had failed to bring about significant change, prompting some Flamingants to push for a law creating separate Flemish and Walloon regiments. Many of the Walloon political leaders saw this as an attempt to divide the Belgian state. After years of proposals, counterproposals, and debate, a law was finally passed in 1913 requiring students in military schools after January 1, 1917, to demonstrate a thorough knowledge of one of the two major national languages and an elementary knowledge of the other.[10] The law came too late and was not strong enough, though, to alter the linguistic composition of the officers' corps during World War I.

More important than the legislative struggle for individual language rights at the beginning of the twentieth century was the continued growth of a sense of Flemish regional distinctiveness and identity. This process began in the nineteenth century with the rise of nationalist feeling, but regional, as opposed to purely linguistic, issues and concerns were becoming somewhat more prominent in the decade and a half prior to World War I. This was both reflected in and encouraged by the growing number of scholarly and literary works devoted to the history and culture of northern Belgium.[11] A symbol of this development was the replacement in 1902 of the famous Flemish literary review, *Van Nu en Straks* (For Now and the Future) by a new periodical entitled *Vlaanderen* (Flanders).[12] Moreover, a few well-

---

[9] Law of May 12, 1910, art. 15. The parent was to declare the language of the child, but school authorities could transfer a child to a school of a different language if the declaration was judged to be incorrect.

[10] Law of July 2, 1913, arts. 1, 2, & 9, *Moniteur Belge*, 83, 212 (July 31, 1913), pp. 5085-5087. The law further required that courses in military rules be offered in both languages (art. 5), a test in Dutch be passed before graduating (art. 6), and notices and orders be given in both French and Dutch.

[11] For example, Frans Van Cauwelaert et al., *Vlaanderen door de Eeuwen Heen* (Amsterdam: Elsevier, 1912).

known Flemish leaders were expressing linguistic concerns in regional terms during this period. Among the most important of these was Lodewijk De Raet, whose comments about the language question at the University of Ghent were noted above. He characterized the Flemish problem in terms of the relative disadvantage of Flanders from an economic and social standpoint. His concerns led him a few years later to write a full exposition of the economic problems of Flanders that was published immediately before the war.[13] A few Flemish radicals went even further, arguing that regional differences should be the basis for the administrative partitioning of Belgium along language lines.[14]

These developments notwithstanding, the orientation of the Flemish movement remained overwhelmingly cultural before World War I, with individual linguistic equality and opportunity dominating the agenda. As Clough pointed out:

> Up to the time of the World War, the leaders of the Flemish movement . . . had no thought of destroying the political unity of Belgium. They were content to limit their demands to a strict equality of Flemish and French in public educational institutions, in the army, in the courts, and in all other governmental services. As is evidenced by this program, the Flemish movement of *ante bellum* days was first of all a struggle for cultural and linguistic equality in Belgium.[15]

Regional socioeconomic concerns such as those of De Raet were shared by only a limited number of Flemings, and calls for a degree of regional political autonomy were even less widespread. Nevertheless, it is important to try to understand the motivations behind this kind of thinking, because it was gaining ground in the years immediately before World War I, setting the stage for the rapid growth in regional-territorial concerns during and after the war.

---

[12] See Clough, *A History of the Flemish Movement*, pp. 109-11.

[13] Lodewijk De Raet, *Vlaanderen's Economische Ontwikkeling* (Brussels: Standaard, n.d.). See also Didaskalos (pseudonym), *De 34 Open Brieven Geschreven door Didaskalos aan M. Helleputte, Minister van Spoorwegen, Posterijnen, Telegrafen en Verschenen in 'De Standaard' van 9 Mei 1907 tot April 1909* (Antwerp: De Standaard, 1909), in which the author draws attention to the imbalance in government spending on Walloons and Flemings.

[14] Their ideas are summarized in Maurice-Pierre Herremans, "Bref historique des tentatives de réforme du régime unitaire en Belgique," *Courrier Hebdomadaire du C.R.I.S.P.*, 135 (January 1, 1962), pp. 4-5.

[15] Clough, *A History of the Flemish Movement in Belgium*, p. 175. See also, M. de Vroede, *The Flemish Movement in Belgium*, trans. from the Dutch by W. Sanders (Antwerp: Kultuurraad voor Vlaanderen, 1975), p. 46.

There can probably be no definitive answer to the question of why some Flemish intellectuals and political leaders began to focus on the regional dimension of the language situation, but a number of factors can be suggested that help to provide an explanation. These are: (1) the desire to broaden the base of support for the Flemish cause throughout the North; (2) the perception and reality of an areal correspondence between economic and linguistic patterns; (3) the recognition that an alternative to individual representation was necessary in order to overcome the disadvantaged position of Flemings in the state; (4) the diffusion of regional-linguistic issues as a consequence of earlier legislative and political developments; (5) the concern over language shift in northern Belgium, particularly in the communes around Brussels; and (6) the regional ideas advanced by a small but vocal group of Walloons reacting to the successes of the Flemish movement.

The base of support for the Flemish movement had grown significantly since its inception shortly after the founding of the Belgian state. Up until 1870 the movement is sometimes characterized as a group of generals with no privates. In the period 1870-1914, however, support for the Flemish cause expanded rapidly, as evidenced by a growing number of Flemish journals and newspapers, which were reaching an ever wider audience during this period.[16] Nevertheless, the movement remained numerically and territorially limited. Support for the organized Flemish cause was weak throughout much of the eastern half of Flanders, and even in western Flanders it was limited largely to the urban middle classes. Moreover, Flanders was not unified from a dialectical or socioeconomic standpoint, and despite the growth of Flemish nationalism, conceptual unity remained limited.

Stressing and encouraging regional continuities was a way of forging a stronger Flemish movement. Efforts to promote Flemish ethnolinguistic consciousness through the symbolism of flags, songs, slogans, and "national" holidays encouraged the development of a sense of regional identity. Flemish leaders hoped to build on this to forge a degree of unity among a still fragmented population. Indeed, stressing shared regional problems and issues could divert attention from some of the social and cultural differences, and functional discontinuities, that continued to divide the Flemings as a people. From a linguistic standpoint there were also encouraging signs for greater regional unity with Flemish linguists finally coming to a consensus on a standard form of speech and writing for Flanders. A debate had raged in intellectual circles since 1830 on this issue

---

[16] See generally, Clough, *ibid.*, pp. 116-119.

impeding any reduction in the dialectical differences that divided the North. By the early twentieth century, however, the new generation of linguists finally came to agree that the language of Flanders must be based on the standard form of Dutch used in Holland.[17] Although it was some time before so-called *algemeen beschaafd Nederlands* (common cultivated Dutch) became widespread, it was already gaining ground rapidly in the years prior to World War I.

From an economic standpoint, although there was little integrity to Flanders as a region, casting issues in terms of oppositions between Flanders and Wallonia provided a means of broadening support for the Flemish cause. The suffrage law of 1893 gave Flemings a greater chance to influence the direction of Belgium's political life if enough Flemings could be mobilized. Appeals based on individual linguistic rights did not have great meaning for many rural, lower-class Flemings who were concerned about the basic issues of livelihood. Bringing these issues into the Flemish movement had obvious implications for the movement's base of support.

At the same time, the economic structure of Belgium itself presented a rationale for regional divisions on the basis of linguistic distribution. The concentration of financial and industrial wealth in part of Wallonia and among the Francophones of Brussels provided a degree of linguistic significance to economic questions, even though the correspondence between language and prosperity was far from complete. The perception of a correlation between linguistic and economic regions was further encouraged when, in the aftermath of the discovery of coal in Limburg in 1901 (figure 4), most of the extraction rights were bought by Walloon or French firms.[18] With the profits destined to leave Flanders, the regional significance of the situation could hardly be ignored. Finally, the prosperity of the Sambre-Meuse valley led to substantial migration from all over Belgium. The number of workers coming from densely populated areas of Flanders was particularly large. Their immersion in a Francophone environment encouraged language shift, thereby undercutting the Flemish goal of linguistic integrity. Once again, the situation had regional-linguistic overtones, since economic opportunity was associated with a change in linguistic context.

In the early twentieth century, the Flemish movement was also confronting mounting frustration over the slow pace of change in the position of the Flemings and their language in Belgium in spite of legislative and political gains. Language laws had been adopted only after

---

[17] See generally, Curtis, "New Perspectives on the History of the Language Problem in Belgium," pp. 252-262.

[18] Clough, *A History of the Flemish Movement*, p. 172.

prolonged struggle and much compromise, and their application was frequently lax. Although this frustration helped to spur the growth of the Flemish movement, suffrage laws and the historical legacy of Francophone dominance in political and social matters effectively meant that the Flemings were not in a position to exercise power in relation to their numbers in the state. Flemish leaders could therefore not expect rapid solutions to their problems in a social and political framework emphasizing individual rights. The alternative of a regional or territorial approach held out the possibility of giving Flemings greater leverage.

In addition, a number of the laws that were in effect embodied distinctions between the language areas of the country, thereby providing limited formal recognition to the Flemish and Walloon parts of Belgium. The necessity of dealing with matters of education, for example, in terms of Flemish/Walloon territorial distinctions promoted thinking about the language areas of Belgium as regional units, as evidenced by the calls for territorial unilingualism in connection with the debate over the school bill of 1914. Moreover, such laws necessitated the specification of Flemish and Walloon communes, thus raising issues of regional extent and delimitation. This is evidenced by maps that began appearing with different interpretations of the language regions of Belgium depending upon the linguistic background of the cartographer.[19]

A long-standing concern of the Flemish movement was the issue of language shift among Flemings. French had made significant inroads in the North for centuries, a situation the Flemish movement sought to reverse through the achievement of linguistic equality for Dutch in the Belgian state. The language laws of the late nineteenth and early twentieth centuries, however, did little to change this process, giving rise to the logical alternative of discouraging language shift through the imposition of regional unilingualism. The situation was brought into particular focus by the changing linguistic character of Brussels and its environs.[20]

---

[19] For example, an early twentieth century map by V. Fris, "Kaart der Taalgrens in België en Noord-Frankrijk" (Ghent: Geirnaert-Vandesteene, 1905), includes Brussels and a large area around Engheim in the Dutch-language area whereas one by Lucien Hochsteyn, "Carte des communes du point de vue de la situation des langues nationales," (Brussels: Institut Cartographique Militaire, 1907), shows Brussels and several surrounding communes as French and only a very small area around Engheim as Dutch. Different ideas about the territorial extent of Flanders and Wallonia are also revealed in the dictionaries of communes that were produced at various times during the twentieth century. For example, Leo De Wachter, *Repertorium van de Vlaamse Gouwen en Gemeenten: Algemeen Gedeelte en Gewesten* (Antwerp: De Sikkel, 1942) includes every commune with even a small Flemish-speaking population. Although beyond the scope of this study, dictionaries of this sort could be used to reconstruct in detail aspects of the evolution of regional conceptions.

The laws calling for the use of Dutch in the courts and schools of the Flemish communes were passed only after compromises had been made involving the settlements being enveloped by the expansion of Brussels. In the Royal Decree of 1891, the communes of Etterbeek, Ixelles/Elsene, Saint-Gilles/Sint-Gillis, Saint-Josse-ten-Noode/Sint-Joost-ten-Node, and Schaerbeek/Schaarbeek were not included among the Flemish communes because they were already regarded as part of the so-called Brussels agglomeration, which was treated as bilingual (see figure 9 in the next chapter).[21] In 1910, five more communes were added to the agglomeration. This meant that an ever widening area around Brussels lost significant legal protections for the Dutch language. Given the dominance of French at the time, it is not surprising that the population of these areas was becoming increasingly frenchified. Although these developments were not frequently challenged before World War I, they did represent a threat to the linguistic objectives of the Flemish movement. As a result, they fueled concern over the inadequacy of laws requiring individual linguistic equality to protect the position of the Dutch language in Belgium, and as territorial developments they encouraged ideas about regional unilingualism as an alternative.

Finally, the reactions of the leaders of the nascent Walloon movement to Flemish successes encouraged regionalism. Interestingly, it was some of these Walloon leaders, much more than their Flemish counterparts, who were promoting the idea of regional administrative autonomy for the language regions. Although the impact of their ideas was limited, they need to be examined because they encouraged the development of regional thinking among some Flemings and Walloons.

Schreurs characterizes 1905-1914 as the period of "la prise de conscience de la Wallonie."[22] A Walloon congress was organized in 1905 to discuss the adoption of the Flemish language as an official language of Belgium along with French.[23] The participants expressed concern that the language laws eroded individual freedom of language choice, although some felt that "Flemish" should not be elevated to a comparable position

---

[20] See generally, Harry Van Velthoven, "De Taalwetgeving en het Probleem Brussel, 1830-1914." *Taal en Sociale Integratie*, 4 (1981), pp. 247-259.

[21] Royal Decree of May 31, 1891, *Moniteur Belge*, 61, 163 (June 12, 1891), pp. 1637-1643. Some of these communes were incorporated into the agglomeration because only those who declared in the 1890 census that they spoke nothing other than Dutch were categorized as Flemish.

[22] "The rise in awareness of Wallonia as an entity." (Author's translation). Fernand Schreurs, *Les congrès du Rassemblement Wallon de 1890 à 1959* (Charleroi: Institut Jules Destrée, 1960), p. 6.

[23] *Congrès Wallon: compte-rendu officiel* (Liège: Mathieu Thône, 1905), p. 7.

with French given the greater international significance of the latter tongue. At both this congress and another one organized in 1906, some participants sought to draw distinctions between Wallonia and Flanders as distinct regional entities.[24] This reflected a desire to encourage a sense of Walloon identity to combat growing Flemish unity and a reaction to fears that the Flamingants were seeking to impose Dutch on the Walloons through bilingual proposals. Belgian unity was not yet brought into question, though.

The situation changed by the time of the next Walloon congress in Liège in 1912. The easy passage of a 1908 law on the language of criminal proceedings in Brussels[25] led to the first Walloon-oriented list of candidates for parliamentary elections in 1910.[26] An organized effort had developed to oppose the conversion of the University of Ghent to a Dutch-language institution,[27] and more attention was being devoted by Francophones to Flanders/Wallonia dichotomies in social and economic matters [28] Concern over the Walloon numerical minority and the legislative successes of the Flemish movement led Emile Jennissen to publish a call for the administrative autonomy of Wallonia in 1911,[29] and the issue was taken up seriously at the 1912 congress. Although it encountered substantial opposition, an ambiguous resolution was passed calling for a separation of Wallonia and Flanders in order to extend Wallonia's independence in relation to the central government and to encourage independent development.[30] This was significant as an early endorsement of regional separatism, but the Walloon movement was so small at the time that its impact was minimal. That a call for regional separatism should have come from the Walloons first is indicative of the intensity of the reaction of a small number of Walloons to Flemish advances. These Walloons

---

[24] Schreurs, *Les congrès du Rassemblement Wallon*, pp. 16-18.

[25] Law of February 22, 1908, *Moniteur Belge*, 78, 61 (March 1, 1908), pp. 1145-1146.

[26] Patrick Van den Berghe and Lode Wils, "De Wet van 22 Februari 1908 op het Taalgebruik in Strafzaken, Bijzonder in het Arrondissement Brussels," *Taal en Sociale Integratie*, 7 (1984), pp. 79-96.

[27] The positions of the organization are set forth in *Union pour la défense de la langue française à l'Université de Gand* (Ghent: A. Vanderwerghe, 1910).

[28] For example, *L'action socialiste: lettres d'Emile Vandervelde "à la Gazette de Bruxelles"* (Ghent: Volksdrukkerij, 1911), pp. 6-8.

[29] Emile Jennissen, ed., *Wallons-Flamands: pour la séparation politique et administrative* (Liège: Charles Desoer, 1911).

[30] Schreurs, *Les congrès du Rassemblement Wallon*, p. 19.

apparently saw regional separatism as the best means of protecting Walloon interests against the growing power of the Flemish majority.

Of far greater national impact than the Walloon congress was a highly publicized letter written a few weeks later to the king by a prominent Walloon, Jules Destrée. Destrée argued that instead of Belgians, there were only Walloons and Flemings, and that the fusion of the two peoples was not desirable.[31] The ideas of Destrée and the more radical participants in the Walloon congress were shared by only a small percentage of Belgians, however. On the Walloon side, they did not result in any general call for the administrative partitioning of the country, although they may have encouraged growing antipathy for the Flemish movement.[32] Certainly in the years immediately before World War I the symbols of Walloon identity multiplied.[33] On the Flemish side, none of the three best-known "answers" to Destrée's letter endorsed federalist ideas.[34] Nevertheless, the Walloon calls for administrative separation were not ignored by the Flemings, and arguably encouraged Flemish leaders such as Leven to begin to question the advisability of a unified Belgian state as World War I approached.[35]

To put the foregoing discussion into perspective, it is important to reiterate that at the outbreak of World War I only a very few Belgians had brought the unity of the state into question. The conceptual division of Belgium into language regions was still limited, and functional and formal divisions were almost nonexistent. The Flemish movement was growing rapidly but remained fundamentally oriented toward language issues, wherasthe organized Walloon movement, being only in its infancy, had a very narrow base of support. Nevertheless, in the early years of the twentieth century the idea of Flanders and Wallonia as significant regional entities within the state was taking hold among a growing number of Belgians. This development was more widespread among Flemings than Walloons, although the ideas of a few Walloons on regional administrative

---

[31] "Excerpts from Jules Destrée's Letter to the King," *Documents Illustrating the History of Belgium*, volume 2, *Modern Belgium from 1830 up to the Present Day* (Brussels: Ministry of Foreign Affairs, External Trade and Cooperation in Development, 1978), doc. 217.

[32] See for example the remarks in *Union des Femmes de Wallonie*, 1 (February 1913), pp. 2-6.

[33] These included a flag of Wallonia (1913) on display at the Musée de la Vie Wallonne in Liège, the resurrection of the journal *La Défense Wallonne* (1913), and the establishment of a museum of Walloon life (1914). *Musée de la Vie Wallonne—Documents* (Liège: Musée de la Vie Wallonne, 1914).

[34] Reprinted in part in W. Houtman, *Vlaamse & Waalse Documenten over Federalisme* (Schepdaal: Het Pennoen, 1963), doc. 5.

[35] Hooger Leven, "De Bestuurlijke Scheiding," reprinted in *ibid.*, doc. 6.

separatism were beginning to take effect even in Flemish circles. As a consequence, Belgium entered the war with a developing sense of regional linguistic dualism.

World War I and the Idea of Flemish Political Autonomy

The German invasion of Belgium in 1914 met with such staunch resistance from the Belgian army that the defense of the country became an international symbol of opposition to the Germans. As the Walloon part of the country was rapidly overrun, the burden of defense fell most heavily on Flemish soldiers who demonstrated a high degree of loyalty to the Belgian state through their valiant efforts. With the inevitable defeat of Belgium in late October 1914, the national government fled to Le Havre and the Germans established a military government for the country. Subordinate functionaries mostly retained their positions, while the upper levels of administration were controlled by Germans.[36]

The majority of Flemings put aside domestic differences and resisted the Germans, but a few opted for collaboration. The former are usually referred to as Pacifists, and the latter as Activists. The Pacifists included many of the supporters and leaders of the Flemish cause before the war, and many were active in opposing the German regime. Some Flamingants, however, became caught up in the nationalist fervor of the time and the upheaval of war, and saw in the German occupation the opportunity to push for greater Flemish rights, and even for self-determination or autonomy. These Activists, many of whom were literary and cultural figures, advocated cooperation with the Germans in an effort to further the Flemish agenda. During the course of the war various Activist groups sought to achieve a variety of objectives ranging from the required use of Dutch at the university level to separation, or even independence, for Flanders.[37] The different views of the Activists and Pacifists on collaboration introduced a deep division into the Flemish movement.

The Germans, following the principle of *divide et impera*, capitalized on Activist cooperation. In an effort to win the help and support of the Activists, the Germans initially acceded to certain of the Flemish demands. The occupation government issued a number of decrees in 1916 to reinforce the 1914 Belgian law requiring that elementary education be in the mother

---

[36] See generally, Adolf Solansky, *German Administration in Belgium* (New York: Columbia Studies, 1928); Ludwig von Köhler, *Die Staatsverwaltung der Besetzten Gebiete Belgiens* (New Haven: Yale University Press, 1927).

[37] These groups are listed, and their platforms summarized, in Clough, *A History of the Flemish Movement*, pp. 188-189.

tongue of the child[38] and the 1878 law on the use of Dutch in administrative affairs.[39] They also extended preexisting requirements on the use of Dutch in Flanders to the operation of the postal service[40] and, in an act foreshadowing things to come, divided the Ministry of Arts and Sciences into separate sections to administer the Flemish and Walloon parts of the country.[41] In addition, a number of communes were added to the list of Flemish communes for purposes of the 1889 law on the language of judicial proceedings, including those of the Brussels metropolitan area.[42] Finally, the Germans sought to reopen the University of Ghent as a Dutch-language institution in 1916. They were only partially successful in this endeavor, however, because most of the prewar faculty refused to participate, many of

---

[38] Decree of February 25, 1916, *Gesetz- und Verordnungsblatt für die Okkupierten Gebiete Belgiens*, 186 (March 9, 1916), pp. 1705-1714, calling for inspection (art. 2) and the end to bilingual classes in Brussels (art. 15); Implementation Regulation of March 18, 1916, *Gesetz- und Verordnungsblatt für die Okkupierten Gebiete Belgiens*, 192 (March 25, 1916), pp. 1791-1796, specifying verification and inspection procedures; Decree of April 22, 1916, *Gesetz- und Verordnungsblatt für die Okkupierten Gebiete Belgiens*, 206 (May 2, 1916), pp. 2056-2059, establishing the presumption in the Flemish part of the country that Dutch was the language of the child (art. 2) and requiring that in schools with pupils speaking more than one tongue, the language of the majority was to be used for instruction; Decree of April 29, 1916, *Gesetz- und Verordnungsblatt für die Okkupierten Gebiete Belgiens*, 208 (May 7, 1916), pp. 2089-2093, specifying communes along the language border to which the decree of February 25, 1916, would apply; Decree of April 30, 1916, *Gesetz- und Verordnungsblatt für die Okkupierten Gebiete Belgiens*, 210 (May 12, 1916), pp. 2113-2117, establishing the same position for French in the Walloon part of the country as for Dutch under the decree of April 22, 1916; Implementation Regulation of May 17, 1916, *Gesetz- und Verordnungsblatt für die Okkupierten Gebiete Belgiens*, 214 (May 23, 1916), pp. 2175-2179, providing for further procedures to verify the language declaration of the head of the family; Decree of June 15, 1916, *Gesetz- und Verordnungsblatt für die Okkupierten Gebiete Belgiens*, 226 (June 24, 1916), pp. 2284-2288, extending the provisions of the decrees of April 22 and April 30, 1916 to state-subsidized nursery schools; Decree of October 4, 1916, *Gesetz- und Verordnungsblatt für die Okkupierten Gebiete Belgiens*, 267 (October 20, 1916), pp. 2833-2837, doing the same for state-subsidized adult education; and Decree of December 13, 1916, *Gesetz- und Verordnungsblatt für die Okkupierten Gebiete Belgiens*, 426 (December 16, 1916), pp. 4893-4894, requiring the use of Dutch in state-subsidized technical schools in Flanders.

[39] Decree of September 2, 1916, *Gesetz- und Verordnungsblatt für die Okkupierten Gebiete Belgiens*, 253 (September 13, 1916), pp. 2617-2622.

[40] Decree of October 15, 1916, *Gesetz- und Verordnungsblatt für die Okkupierten Gebiete Belgiens*, 272 (November 2, 1916), pp. 2897-2902.

[41] Decree of October 25, 1916, *Gesetz- und Verordnungsblatt für die Okkupierten Gebiete Belgiens*, 273 (November 5, 1916), pp. 2930-2931; and Decree of December 13, 1916, *Gesetz- und Verordnungsblatt für die Okkupierten Gebiete Belgiens*, 288 (December 15, 1916), pp. 3054-3056.

[42] Decree of November 15, 1916, *Gesetz- und Verordnungsblatt für die Okkupierten Gebiete Belgiens*, 279 (November 21, 1916), pp. 2978-2979.

the replacements were not Belgians, and the number of students was very low.[43]

Although these initiatives encouraged the Flemish Activists, some of the more extreme elements wanted to push for administrative partitioning or independence for Flanders so as to be in a better position to obtain national autonomy after Germany's presumed victory. Out of this grew the so-called Council of Flanders, a group of intellectuals and officials selected at an Activist gathering in February 1917. The Germans recognized this council as the representative organ of the Flemish people despite the refusal of most Belgian administrative authorities and Flemings to recognize its authority.[44] Although the Council saw itself as an independent body, the Germans exercised a great deal of control over it and regarded it as a tool for achieving German objectives in the region.[45] The Germans went along with a council proposal to institute an administrative division of the country along language lines that many of the Activists hoped would be a step toward independence. It was initiated by decree on March 17, 1917, and Namur and Brussels were declared the capitals of Wallonia and Flanders, respectively.[46] In applying the law, separate administrations were set up for Wallonia and Flanders, and certain provincial and communal boundaries were adjusted to conform to the administrative regions.[47]

In response to the administrative division, the Council of Flanders adopted a statement of allegiance to Germany in June 1917.[48] At the same time, Council members began to push for greater autonomy for Flanders. The Germans appeared to cooperate with these initiatives, but when the Council issued a call for the complete independence of Flanders on March 28, 1918, the Germans suppressed its publication.[49] Council members lodged

---

[43] See generally, *De Vlaamsche Hoogeschool te Gent, 1916-1918* (Ghent: Plantijn, 1919).

[44] See generally, *Les archives du Conseil de Flandre (Raad van Vlaanderen)* (Brussels: Ligue Nationale pour l'Unité Belge, 1928).

[45] See the correspondence between the governor general of Belgium and Hindenburg quoted in *ibid.*, p. xxiv.

[46] Decree of March 21, 1917, *Gesetz- und Verordnungsblatt für die Okkupierten Gebiete Belgiens*, 324 (March 21, 1917), pp. 3457-3458.

[47] Decree of April 13, 1917, *Gesetz- und Verordnungsblatt für die Okkupierten Gebiete Belgiens*, 335 (April 19, 1917), pp. 3597-3598; and Decree of August 9, 1917, *Gesetz- und Verordnungsblatt für Wallonien*, 10 (February 4, 1918), pp. 49-50 and *Gesetz- und Verordnungsblatt für Flandern*, 11 (February 4, 1918), pp. 95-96.

[48] *Les archives du Conseil de Flandre*, p. xxx.

[49] *Ibid.*, p. li.

a protest and concern developed over German intentions. In an effort to placate the Activists, in June 1918 the Germans enacted additional ordinances strengthening the position of Dutch in Flanders.[50] Further efforts on either side were soon overshadowed, however, by the impending defeat of Germany.

At the same time that the Activists were pushing for the political reorganization of Belgium, a movement was developing among Flemings in the Belgian army. Although it may have been stimulated indirectly by Activist propaganda, it essentially developed as a reaction to the lowly position of the Dutch-speaking Flemish soldier in the military hierarchy. Even though the vast majority of soldiers were Flemings, commands were given in French, signs were in French, and promotion was extremely difficult for Dutch speakers. Despite the efforts of some Belgian authorities to encourage commanders to learn and use Dutch, the situation did not improve, and in 1916 a loose organization of radical Flemish soldiers was formed. The organization evolved into the so-called Front Party (Frontpartij), and promoted feelings of solidarity among Dutch-speaking soldiers.

The significance of these developments for the growth of regional consciousness becomes evident when one considers that the army brought together Flemings from all over northern Belgium who came to realize that their language put them at a common disadvantage. Soldiers from Limburg and West Flanders saw themselves as fundamentally united by a shared problem. It is hardly surprising, then, that Flanders as a region should have become an important symbol for the so-called Frontists, as attested to by the inscription placed on the graves of the Flemish dead, "*Alles voor Vlaanderen, Vlaanderen voor Kristus*" (All for Flanders, Flanders for Christ). Years of suffering together during the war convinced some of the soldiers that Belgian integration and unity meant nothing other than Francophone dominance. This, combined with the news of Activist initiatives that filtered through to the front, led the Front Party, in a platform issued in 1917, to call for the administrative partitioning of Belgium and the division of the army into Flemish and Walloon units.[51]

---

[50] Ordinance of June 4, 1918, *Gesetz- und Verordnungsblatt für Flandern*, 65 (July 6, 1918), pp. 615-620, prohibiting instruction in French in Flemish schools below the fourth level, and after that limiting it to no more than three hours per week (art. 8); Ordinance of June 4, 1918, *Gesetz- und Verordnungsblatt für Flandern*, 65 (July 6, 1918), pp. 620-622, requiring the use of Dutch in all middle schools in Flanders; and Ordinance of June 4, 1918, *Gesetz- und Verordnungsblatt für Flandern*, 65 (July 6, 1918), pp. 622-623, stipulating that after January 1, 1919, no one could become a public employee without passing an examination in Dutch.

[51] Rudiger [pseudonym for A. Wullus], *Flamenpolitik* (Brussels: Rossel, 1922).

It is often said that war tends to precipitate change at an accelerated rate, and this was certainly true with respect to the Flemish movement during World War I. In four years the Flemish cause evolved from a more or less unified and growing cultural struggle stressing individual linguistic rights to a movement divided between Pacifists, who, while maintaining their cultural and linguistic concerns, thought it inappropriate to press Flemish demands in an occupied state, Frontists, who recast the focus of the movement in political and regional terms, and Activists, who sought to exploit the situation to achieve territorial and national goals. The emphasis of both Frontists and Activist leaders on regional autonomy for Flanders marks the beginning of organized territoriality in the Flemish movement. It did not spring up by chance during the war, however. It was during and immediately after the war that nationalist ideas were at their apogee in Europe. Moreover, as has been argued, the seeds of Flemish regionalism were sown before 1914. The factors promoting the rise of regional-territorial concerns discussed earlier in this chapter simply became more compelling in the social and political upheavals accompanying military struggle and occupation, and in light of the uncertainty over Belgium's and Europe's future.

Regionalist ideas were not embraced by all Flemings. The strong opposition of most Flemings to the Germans and, hence, to the cooperative attitude of the Activists, meant that ideas about Flanders' independence were rejected by the vast majority. Even the more moderate Frontist regionalist ideas were regarded with some suspicion by many Flemings. In addition, the impact of the administrative division of Belgium was limited as it was dismantled immediately after the defeat of the Germans. Nevertheless, the influence of wartime developments on linguistic regionalism should not be underestimated. The enduring legacy of this period was "to put political independence on the agenda of the Flemish movement for the first time."[52] Activist and German propaganda had brought the issues of regional linguistic autonomy before most Belgians, and the conceptual and functional consequences, results of the short-lived but far-reaching formal administrative division of the country, could not easily be reversed.

---

[52] Kenneth D. McRae, *Conflict and Compromise in Multilingual Societies: Belgium* (Waterloo, Ontario: Wilfrid Laurier University Press, 1986), p. 27.

Postwar Reactions

Despite the hostility in Belgium immediately after the war toward the Activists and other collaborators with the Germans, the king took a conciliatory attitude to the more moderate Flemish grievances in his victory speech to the Belgian people. In so doing, he was probably seeking both to avoid the internal upheaval that had beset other parts of Europe and to direct attention away from Flemish regionalism, which had blossomed during the war among Activists and Frontists. He essentially took the side of the Pacifist Flemings, calling for cooperation in bringing about strict equality of language and promising to propose to the parliament the creation of a Flemish university at Ghent.[53] In a show of good faith, the king also established a commission to study the problems relating to the use of French and Dutch in Belgium.[54]

The wave of patriotism that swept Belgium following victory over the Germans was accompanied, however, by a severe, even harsh repression of unpatriotic behavior.[55] Walloons played a significant role in this process. A high priority was the prosecution of the Activist collaborators who had remained in Belgium after the war. The trials were dramatic and highly publicized, and the accompanying rhetoric often went beyond the allegations against the accused to sweeping derogatory comments about the Flemish cause in general. As a result, even though the majority of Flemings had been Pacifists during the war, "the repression made them feel as if it was their ethnic status as Flemings that was under attack, not just the actions of a few collaborators."[56] Passions were also aroused by the harsh sentences. These included the death sentence, although no one was actually executed. The trial of the prominent Flemish Activist August Borms was particularly dramatic: As his death sentence was being read, Borms continued to intone "Long live Flanders, long live Flanders,"[57] sentiments that struck a chord with many Flemings. Thus, although the postwar suppression of the Activists was a setback in some ways for the Flemish

---

[53] *Pasinomie*, 5th ser., 9 (November 22, 1918), p. 104.

[54] Royal decree of October 15, 1918, *Moniteur Belge*, 88, 286-292 (October 13-19, 1918), pp. 833-836.

[55] James A. Dunn, Jr. "Social Cleavage, Party Systems and Political Integration: A Comparison of the Belgian and Swiss Experiences" (Ph.D. dissertation, University of Pennsylvania, 1970), p. 111.

[56] *Ibid*.

[57] Clough, *A History of the Flemish Movement in Belgium*, p. 221.

movement, it intensified antagonism between the language groups and consequently promoted regional linguistic distinctions.[58]

The Flemish movement was fundamentally divided after the war between those who sought solutions to Flemish grievances in the context of the unitary Belgian state (Minimalists) and those who thought that Flemish problems could not be solved short of regional autonomy or independence (Maximalists). It was the Maximalists who carried forth Frontist and Activist regional ideas into postwar Belgium, and although their base of support was relatively narrow, they had a decisive impact on growing regionalism in Belgium. The most important Maximalist group was a successor to the wartime Front Party. Its platform called for: (1) autonomy for Flanders in the realms of administration, government, education, and justice; (2) a federalist structure for Belgium; and (3) the reorganization of the military along regional linguistic lines.[59] In other matters, Maximalist concerns did not differ greatly from those of the Minimalists; they included such issues as amnesty for political prisoners, the abolition of the teaching of French in the elementary schools of Flanders, and the conversion of the University of Ghent to a Dutch-language institution. The Front party also fielded candidates for political office. Although the party never sent a large delegation to Parliament, in the ten years after the war electoral support grew steadily.[60]

Branded by the stigma of collaboration attached to the radical wing of the Flemish movement, the Maximalists remained marginal in Belgian politics for over a decade. It was therefore left largely to the Minimalists to advance the Flemish cause. The Minimalists, for their part, campaigned for education in the mother tongue of all children, instruction in Dutch at the University of Ghent, and stronger laws on the use of Dutch in administration and government in the Flemish provinces.[61] Progress in realizing even the more moderate Minimalist demands was slow due to postwar preoccupation with rebuilding and the prosecution of collaborators, although the divisions that had arisen in the Flemish movement also

---

[58] For a study of the Flemish movement between the wars, see A. W. Willemsen, *Het Vlaams-Nationalisme: De Geschiedenis van de Jaren 1914-1940* (Utrecht: Ambo, 2nd ed., 1969).

[59] Platform of the Front movement as it appeared in *Ons Vaderland*, March 4, 1919, reprinted in translation in Pro Flandria Servanda, *Flanders' Right and Claim for Autonomy Formulated, Explained, Justified* (The Hague: Martinus Nijhoff, 1920), p. ix.

[60] See generally, Clough, *A History of the Flemish Movement in Belgium*, pp. 230-231.

[61] See generally, Lode Claes, "Le mouvement flamand et la réforme de l'état," In *La réforme de l'état* (Brussels: Centre d'Etudes pour la Réforme de l'Etat, 1937), pp. 365-368.

impeded progress. Many Flemings felt that they had shouldered more than their fair share of the burden for the defense of Belgium, and that they were entitled to complete equality of language and opportunity. Impatience over the rate of change mounted, particularly among young students at the universities of Ghent and Louvain, inducing the Flemish leader Van Cauwelaert to write to a friend in 1919 that "the unwillingness of the government to give the Flemings their rights drives our intellectual youth more and more toward the notion of [administrative] separation."[62] Thus, the slowness of change and inadequate enforcement of the existing laws encouraged the development of more radical ideas.

In 1919 the one-man-one-vote principle was instituted in Belgium, giving the Flemings the potential to exercise greater control in politics as a consequence of their numerical superiority.[63] Some three percent of the Flemish population spoke French, but that was not enough to offset a substantial population difference between Flanders and Wallonia.[64] It was some time before numerical superiority turned into greater political power, however, because of the historical legacy of entrenched Francophone control of political and economic institutions, the ignorance of or apathy toward the organized Flemish cause on the part of many Flemings, and the continued importance of other cleavages, particularly among Catholics, Liberals, and Socialists, in determining voting patterns. Nevertheless, the suffrage change was in part responsible for the first postwar legislative success of the Flemish movement, namely the law on the use of languages in administrative affairs.[65] The continued dominance of French in public affairs throughout northern Belgium prompted the Minimalists to focus on this issue and to take a strongly territorial approach. Reflecting the growing regionalist ideology of the time, they sought a division of Belgium into two linguistic sections for administrative purposes, although the moderates did

---

[62] Quoted in Curtis, "New Perspectives on the History of the Language Problem in Belgium," pp. 309-310.

[63] Pierre Maroy, "L'évolution de la législation linguistique belge," *Revue du Droit Public et de la Science Politique*, 82, 3 (May-June 1966), pp. 449-501; Val R. Lorwin, "Linguistic Pluralism and Political Tension in Modern Belgium," *Canadian Journal of History*, 5, 1 (March 1970), p. 7.

[64] According to the 1920 census, the predominantly Flemish communes and districts, excluding the Brussels district, contained 3,994,024 people or 53.9 percent of the Belgian population, whereas the Walloon provinces contained 2,655,440 people or 35.9 percent of the Belgian population. Calculated from *Recensement Général du 31 Décembre 1920: Population*, vol. 1 (Brussels: Ministère de l'Intérieur et de l'Hygiène, 1926), pp. 44-45. The three percent figure for the French-speaking population of Flanders is an estimate of Dirk Wilmars for the early twentieth century. Dirk Wilmars, *Le problème belge: la minorité francophone en Flandre*, trans. from the Dutch by Edmond Knaeps (Antwerp: Scriptoria, 1968), p. 201.

not insist on the total exclusion of French from Flanders. Long, acrimonious, and highly publicized debates took place over an extended period of time, with some Walloons accusing the Flemings of wanting to split the country in two. The Flemish victory eventually came in 1921.

The law stipulated that in the provinces of Antwerp, West Flanders, East Flanders, and Limburg, as well as in the districts of Leuven and Brussels with the exception of the communes of metropolitan Brussels, Dutch was to be used in state, provincial, and communal administrative affairs, with an exemption given to personnel hired before January 1, 1920. Provincial and municipal councils were free to add French, but not to substitute it for Dutch. Corresponding provisions covered the use of French in the Walloon part of the country. Since district and provincial boundaries did not correspond entirely to language lines, the law allowed that if the majority of inhabitants in a municipality spoke a different language from that of the linguistic part of the country to which it was assigned, the communal council could decide which of the national languages to use. A bilingual regime was established for Brussels, and in the central administration the use of language was to depend on the part of the country from which the interested parties came. Despite the call of some Flemings for the inclusion of sanctions for violations, the strong opposition of many Walloons and the lack of support from some Minimalists led to the abandonment of this idea.[66] Nevertheless, the law was the first important formal recognition of language areas as a basis for regional distinctions within the state and, as such, represents the growing focus of political leaders on the territorial dimension of language practices.

Encouraged by their success, Flamingants turned their attention once again to the issue of language usage at the University of Ghent. The Maximalists sought the immediate transformation of the university to a totally Dutch-language institution, the Minimalists were willing to consider a more gradual change, and many Walloons rejected the idea altogether. After much debate, a compromise was reached in 1923 that satisfied almost no one but that was all that the Flemish politicians could realistically achieve at the time. It called for Dutch to be the language of administration at the university, and in most of its sections, for students to be able to choose a program in which two-thirds of the courses were in one of the national languages, and one-third in the other.[67] It was not until 1930 that the

---

[65] Law of July 31, 1921, *Moniteur Belge*, 91, 224 (August 12, 1921), pp. 6566-6568.

[66] Clough, *A History of the Flemish Movement in Belgium*, p. 234.

[67] Law of July 31, 1923, *Moniteur Belge*, 93, 213 (August 1, 1923), pp. 3712-3714.

Flemish movement managed to overcome internal divisions and postwar suppression to achieve the complete transformation of Ghent into a Dutch-language institution.[68]

With the exception of two royal decrees calling for the translation of laws and the Constitution into Dutch,[69] the only other legislative success of the Flemish movement in the 1920s concerned language use in the army. A bill was introduced in 1928 calling for separate regiments for the two language groups. Despite dire forebodings on the part of some Walloons, the Minimalists and Maximalists united behind the bill, and it was passed later in the year.[70] This law, together with the administrative law of 1921, represented the only significant successes for the Flemish movement during the entire decade after World War I. At the same time, awareness of the disadvantaged position of Dutch speakers was growing and Flemish propaganda was reaching an ever wider audience. It is consequently not surprising that the Flemish movement--"in both its cultural and political manifestations--finally struck root in the masses during this period."[71] The movement was particularly strong at the universities, and as students graduated and entered professional positions it became more widely diffused.[72]

As the legislative initiatives discussed above indicate, in the postwar period the objectives of a significant element of the Flemish movement went well beyond linguistic concerns. They encompassed social stratification, economic well-being, educational reform, and Flemish solidarity, issues that were fundamentally tied to regional integrity. Because of postwar repression, it took more than a decade for the Flemish movement to become a major political force in Belgium in spite of suffrage changes. The movement that emerged in the late 1920s, however, was stronger and more radical in its objectives as a result of shared suffering during the war, years of repression and frustration thereafter, and the growing spotlight on ethnoregional, as opposed to purely linguistic, issues.

Before analyzing this development, a few words need to be said about Walloon sentiments during the postwar years. As mentioned already,

---

[68] Law of April 5, 1930, *Moniteur Belge*, 100, 106 (April 16, 1930), pp. 1928-1929.

[69] Royal Decree of September 18, 1923, *Moniteur Belge*, 93, 271 (September 28, 1923), pp. 4814-4815; Royal Decree of November 25, 1925, *Moniteur Belge*, 95, 353 (December 19, 1925), p. 6362.

[70] Law of November 7, 1928, *Moniteur Belge*, 98, 315 (November 10, 1928), pp. 4638-4641.

[71] Curtis, "New Perspectives on the History of the Language Problem in Belgium," p. 316.

[72] Lode Claes, "Le mouvement flamand entre le politique, l'économique et le culturel," *Res Publica* 15, 2 (1973), p. 222.

euphoria over the defeat of Germany led to a surge of patriotism among the Walloons, and the federalist rhetoric of the prewar militants died down for a while.[73] Nonetheless, concern was growing among a small segment of the Walloon population over the implications of their numerical minority within the state and the growing need to demonstrate knowledge of both French and Dutch to obtain certain positions in the government and the military. As a British map of Belgium's languages prepared during the war indicates, there were very few Dutch speakers in the South,[74] and it is well known that few Walloons were competent in Dutch as a second language. In contrast, a high percentage of educated Flemings knew French, putting them in an advantageous position to compete for bilingual positions. By the middle of the 1920s the Ligue d'Action Wallonne had begun organizing annual meetings at which concerns over these issues were expressed. Although the meetings were poorly attended, a few Walloon militants argued that a federalist structure was the only way to protect Walloon interests.[75] For these people, federalism was a way of ensuring unilingualism in the South and an equal share of central power.

The Push for Territorial Unilingualism

By the late 1920s, the Flemish movement was emerging from the period of postwar suppression as a major political force in Belgium. This was dramatically demonstrated in an Antwerp by-election for the House of Representatives in 1928 in which the former Activist collaborator Borms, who was still in prison in Leuven, received two-thirds of the vote.[76] The emergence of regional issues and concerns and the growth of Flemish ethnolinguistic identity during and after World War I had irrevocably changed the character of the movement. The new generation of Flemish leaders, many of whom had been involved in the student movements at the universities in Ghent and Leuven after the war, were no longer content to press for individual language rights. Instead, they saw some degree of

---

[73] See generally, Schreurs, *Les congrès du Rassemblement Wallon*, pp. 22-23.

[74] *A Manual of Belgium and the Adjoining Territories: Atlas* (London: Naval Staff Intelligence Division, 1918), map 4.

[75] Herremans, "Bref historique des tentatives de réforme," p. 5.

[76] Described in Manu Ruys, *The Flemings: A People on the Move, A Nation in Being*, trans. from the Dutch by Henri Schoup (Tielt: Lannoo, 2nd ed., 1981), p. 90. Borms was not allowed to take his seat in Parliament, but his election precipitated the enactment of an amnesty bill in 1929 leading to Borms' release from prison and allowing the Activists who had fled to Holland to return. The bill did not, however, call for the return of confiscated property or the restoration of civil rights for those who had been condemned to more than ten years in prison.

regional autonomy for Flanders as the best hope for achieving linguistic equality within the state. Many argued for territorial unilingualism while others sought a more sweeping restructuring of the Belgian state along language lines.

The orientation of this new generation of Flemish leaders shows how far regional thinking had evolved during and immediately after World War I. The factors identified earlier in this chapter as being influential in the development of limited regional-linguistic ideas before the war had become even more compelling in the ensuing decade and a half. Initiatives grounded in the liberal notion of individual language choice in Flanders had made so few inroads into Francophone dominance--despite suffrage changes and increased support for the Flemish movement--that frustration and thoughts about alternative approaches were almost inevitable. In addition, language shift persisted unabated in the area around Brussels, and ineffective enforcement of earlier laws continued to raise regional-linguistic concerns. At the same time, the new law on language use by public officials and agencies added another element of differentiation between the North and the South. Finally, much of Flanders remained economically disadvantaged in comparison to parts of Wallonia, and Walloons or Francophone Flemings continued to control much of the investment in the North.

A number of other developments in the early twentieth century had served to promote the diffusion of regional ideas. These included: (1) the growth of regional consciousness, and even ideas about regional autonomy, among Flemish soldiers as a result of contact and shared suffering in the army; (2) the development of opportunistic regional ideas among Activists based on assumptions of a German victory and a dissolution of the Belgian state; (3) German efforts to encourage regional division in Belgium, including the imposition of an administrative partition of the state along language lines; (4) regional oppositions that arose out of the postwar suppression of collaborators; (5) the increase in unilingualism in Flanders as a consequence of both the gradual spread of standardized Dutch and the success of the earlier language laws in reducing the use of French; and (6) the significant propaganda and publicity surrounding Frontist, Activist, and Maximalist regionalist ideas. Irrespective of one's attitude toward regional linguistic, cultural, administrative, or political autonomy, by the late 1920s virtually every Belgian had been confronted with the regional ideas regarding the so-called language question, and discourse increasingly reflected regional conceptualizations.

In 1929 a group of Flemish nationalists who were convinced that the resolution of the Flemish question lay in regional autonomy or independence for Flanders sought to develop a specific program to bring before the Flemish electorate. They organized the Vlaamsch Studiecomité voor Politieke, Sociale en Cultureele Aangelegenheden, which investigated various options for Flanders including federalism within the Belgian state, independence, and even union with the Netherlands. Their report, issued in 1930, adopted a federalist stance.[77] Shortly thereafter, Herman Vos introduced a proposal that was supported by members of the Flemish nationalist party to amend the Constitution and restructure the Belgian state as a federation.[78] The project was immediately attacked from all sides, however, and it soon became clear that it had little chance of success.

In what was probably a pragmatic move, Vos soon retreated from his federalist proposal and introduced a new bill calling for complete regional unilingualism in provincial and local government.[79] The accompanying report by the Flemish moderate Van Cauwelaert noted that the 1921 law, by allowing French to be used as the second language in Flanders, had failed to bring about much change, particularly since "public administrators at all levels had not shown the desire to apply the law in good faith."[80] Van Cauwelaert called for complete regional administrative unilingualism, reorganization of the central administration in accordance with the country's linguistic duality, equal respect for the national languages, and obligatory bilingualism of public officials in Brussels. Members of the section reviewing the bill gave some consideration to the idea of changing the provincial boundaries to reflect the language border, but the majority rejected the idea as being too complex and sensitive to treat at the time.

The bill perhaps would not have passed if support had not come from some Walloons. Yet Walloon sentiment for regional unilingualism was definitely growing during this period. Faced with a gradual decline in Walloon political power, a lower population growth rate in the South, and a slow but growing shift in the economic center of gravity to the North, some Walloons were beginning to develop a minority complex. The Borms

---

[77] Vlaamsch Studiecomité voor Politieke, Sociale en Cultureele Aangelegenheden, "Een Politiek Statuut voor Vlaanderen," excerpt from *Federalisme en Groot-Nederland*, reprinted in W. Houtman, *Vlaamse & Waalse Documenten over Federalisme* (Schepdaal: Het Pennoen, 1963), doc. 17.

[78] "Bondsstatuut van het Vereenigd Koninkrijk Vlaanderen en Wallonie," *Documents Parlementaires—Chambre—1930-1931*, March 25, 1931, doc. 177. Interestingly, the proposal dealt with the problem of Brussels by simply making it a part of Flanders. In fact, Brussels was designated as the capital of Flanders and the bill provided for the choice of a capital for Wallonia.

[79] *Documents Parlementaires—Chambre—1931-1932* (December 16, 1931), doc. 67.

[80] *Ibid.*, p. 2. Violations had been made easier by the lack of sanctions in the 1921 law.

election had made it clear that the Flemish movement was no longer a few fanatics seeking to make trouble and, presented with the choice of universal bilingualism or territorial unilingualism, many Walloons opted for the latter. Knowledge of Dutch was considerably more limited in the South than was knowledge of French in the North, and French was widely regarded as the more important international language. Consequently, many Walloons strongly opposed any suggestion that the study of Dutch be required in Wallonia or that bilingualism be requisite for public office.

At a Walloon congress organized by radical Walloons in Liège in 1930, discussion centered around ways of encouraging the growth of ethnolinguistic identity among Walloons and of achieving cultural autonomy for Flemings and Walloons.[81] The congress became an annual event, and at the meeting the following year, the participants went so far as to adopt a proposal for the reorganization of Belgium as a federation consisting of Flanders, Brussels, and Wallonia.[82] Although the views coming out of the Walloon congresses in the 1930s were not shared by many of the inhabitants of southern Belgium at the time, they reflect a growing feeling among some Walloons that a form of territorial unilingualism would be necessary to preserve their cultural and linguistic integrity. In fact, the publicity surrounding the congresses helped to bring the ideas of the participants before the Walloon public.

With some support coming from both Flemish and Walloon elements for a territorial approach to language legislation, the Vos proposal was taken seriously. It emerged from Parliament in June 1932 as the first law to call for complete administrative unilingualism in the Flemish and Walloon parts of Belgium.[83] As for the central government, the law required that official communications be in both languages, that public officials respond to matters in the language in which they were addressed, that acts and certificates be in the language requested by the interested party, and that personnel be grouped by language so that departments, if not individuals, would be bilingual. The final version of the bill reflected the inability of the legislators to reach a decision on the Brussels situation. Although it was agreed that external services were to be bilingual, the bill

---

[81] *Congrès de Concentration Wallonne—Liège—27-28 septembre 1930: compte-rendu officiel* (Liège: Wallon Toujours, 1930).

[82] Schreurs, *Les congrès du Rassemblement Wallon*, pp. 24-25.

[83] Law of June 28, 1932, *Moniteur Belge*, 102, 181 (June 29, 1932), pp. 3577-3580. The language regions were delimited along provincial and district lines exactly as they had been specified in the 1921 law. Special provisions for communes which did not share the language of the region are described below.

simply provided that decisions about internal language use were "to take into consideration local circumstances."[84] Finally, with regard to the communes around Brussels and near the language border, if the decennial census indicated that the majority of the population spoke a different language from that of the region to which the commune was assigned, public officials could use the language of the majority. For those communes in which the census showed that 30 percent or more of the inhabitants spoke a language other than that of the majority, public officials were required to issue notices to the public in both national languages and to provide certain services in the minority language. The law did not provide any sanctions for violations, however, an apparent concession of the bill's proponents to ensure its adoption.

The principle of territorial unilingualism was furthered by a law enacted only two weeks later stipulating that the regional language be the official language of education.[85] Outside of Brussels, children could begin learning the second national language only after the fifth year of schooling, and then only for four hours per week. The children of parents belonging to linguistic minorities in Flanders and Wallonia could begin school in their mother tongue, but they were to enroll in "transmutation" classes so as to learn the regional language as fast as possible and continue their education in that language. In Brussels and the communes of the language border, children were to be educated in their mother tongue as declared by the head of the family and verified by school administrators. Instruction in the other national language, however, was obligatory beginning with the third year of school. Once again, the law failed to provide for sanctions in the event of violations.

Earlier laws concerning the language of education had recognized the principle that the regional language should be the basis for education in the Flemish and Walloon parts of Belgium. As Elizabeth Swing points out: "What was new was the idea that geography should determine the language of education for *everyone*, even the children of linguistic minorities."[86] The adoption of the two 1932 laws on language in public affairs and

---

[84] *Ibid.*, art. 1, §4.

[85] Law of July 14, 1932, *Moniteur Belge*, 102, 216 (August 3, 1932), pp. 4121-4125.

[86] Elizabeth S. Swing, *Bilingualism and Linguistic Segregation in the Schools of Brussels* (Quebec: Centre International de Recherche sur le Bilinguisme, 1980), p. 72. Although almost all state schools in Wallonia were conducted in French anyway, in Flanders, as of 1929, 14 localities and 252 official classes had been organized in French at the primary school level, and this did not even include the many private schools in French. Léon Bauwens, *Régime linguistique de l'enseignement primaire et de l'enseignement moyen* (Brussels: Edition Universelle, 1933), p. 7.

education meant that for the first time—outside of Brussels and the communes of the language border—where people lived rather than individual preference determined language practice. Lorwin notes that by taking this approach, "Belgium accepted in language an analogue to what the Peace of Augsburg had given the Germanies in religion in 1555. The principle was that of territorial, rather than personal, choice."[87] One hundred years after the founding of the Belgian state, regional-linguistic ideology was replacing liberal ideology in state policy toward the so-called language problem.

The Implications of Territorial Unilingualism

The 1932 laws failed to bring about the degree of change expected by many. Violations took place at all administrative levels and in the absence of sanctions little could be done. The education law in particular was widely evaded. In Flanders, many French and Dutch-speaking parents sent their children to French-language schools located immediately to the south of the language boundary or to nonsubsidized French private schools, while the "transmutation" classes, which were supposed to prepare children for education in the regional language, were often turned into a means of avoiding instruction in the regional language.[88] In Brussels, although children theoretically had the right to choose between Dutch and French language programs, a number of the communes did not even have Dutch-speaking schools. Finally, declarations as to the mother tongue of children in Brussels were apparently often false, and there was no effective inspection or supervision of the declarations.[89]

The 1932 laws were nevertheless significant in framing the language question in regional terms. In the wake of their enactment, proposals blossomed among certain Flemish and Walloon intellectuals and politicians for the territorial restructuring of the Belgian state along language lines. The Flemish movement was seriously divided during the 1930s in its

---

[87] Val R. Lorwin, "Linguistic Pluralism and Political Tension in Modern Belgium," *Canadian Journal of History*, 5, 1 (March 1970), p. 13.

[88] See generally, M. de Baere, "L'application de la loi du 14 juillet 1932 dans les établissements d'enseignement moyen et son influence sur les connaissances linguistiques des élèves," Centre de Recherche pour la Solution Nationale des Problèmes Sociaux, Politiques et Juridiques en Régions Wallonnes et Flamandes, documents in the collection of the Bibliothèque Royale Albert 1er, Brussels, Belgium, doc. no. 18.

[89] Centre de Recherche pour la Solution Nationale des Problèmes Sociaux, Politiques et Juridiques des Diverses Régions du Pays, *Rapport Final*, issued as *Documents Parlementaires—Chambre—1957-1958*, April 24, 1958, doc. 940, section culturelle, p. 11.

approach to this question. At the more extreme end, some Flemings argued for federalism, while a few even contemplated a union of Flanders with the Netherlands.[90] In the period from 1930-1938, federalist ideas were promoted by a new Flemish nationalist party, the Vlaamsch Nationaal Verbond, and by a group of Flemish professors at the University of Louvain in the influential journal *Nieuw Vlaanderen*.[91] Although these ideas were not widely shared, they helped stimulate more moderate proposals for other laws addressing the national linguistic and territorial duality.

During the 1930s a growing number of Flemings were calling for some degree of cultural or administrative autonomy for the language regions. In the summer of 1935 the widely circulated Dutch-language newspaper, *De Standaard*, published a battle plan for the future that included official determination of the language border and the extent of the Brussels bilingual area and the splitting of government services along language lines.[92] A Flemish faction of the Catholic party that had broken away from the parent party to form the *Katholieke Vlaamsche Volkspartij* was behind the plan, and at a congress in 1937 it voted to divide Belgian territory for political and administrative purposes.[93] At a congress in 1937 Flemish socialists also endorsed a plan to bring about conformity between administrative and linguistic borders.[94] The growing attitude among Flemings during this period is best expressed in the popular slogan of the time, "De Vlaming meester in eigen huis" (Flemings [should be] masters in their own house), and by the highly publicized and successful actions of the Flemish teacher Flor Grammens. Frustrated over the growth of French influence along the language border, he began a one-man campaign to obliterate bilingual signs with paint and brush. Territoriality as a strategy for achieving Flemish demands was clearly gaining momentum.

Factionalism among Flemings, the economic crisis in Belgium, and the growth of a fascist element among some Flemish nationalists that alienated them from the masses meant that there were few additional

---

[90] There was nothing new about this suggestion, although it had never been widely supported by Flemings. Moreover, proponents of union with the Netherlands received little encouragement from the Dutch government or people, probably due to religious differences and the separate development of Flanders and the Netherlands since the sixteenth century, outside of the brief period of union after the defeat of Napoleon.

[91] See generally, Lode Claes, "Le mouvement flamand et la réforme de l'état," pp. 376-381.

[92] Reprinted in Ruys, *The Flemings*, p. 100.

[93] Claes, "Le mouvement flamand et la réforme de l'état," p. 384.

[94] *Documents Parlementaires—Chambre—1961-1962*, December 21, 1961, doc. 194-7, p. 18.

legislative successes during the 1930s. Exceptions were two essentially nonterritorial laws: a 1935 law calling for sanctions if the language of the parties was not used in judicial proceedings,[95] and a 1938 law reinforcing the principle of dual unilingualism in the armed forces.[96] In the context of the developing discussion over the regional restructuring of the Belgian state, however, new issues were coming to the fore. These included the role of Brussels, the advisability of formal recognition of a fixed language border, and the desirability of economic integration. Many of the Flemish federalist and regionalist proposals were aimed at decreasing the influence of an increasingly frenchified Brussels in Belgium, and a number dealt with the Brussels situation by simply incorporating it into Flanders. The issue of fixing the language border was of concern because under the laws of the 1930s there was nothing to prevent the continued expansion of French influence. Nevertheless, a generally acceptable agreement on the language border was clearly a long way off. With regard to the economic question, most Flemish proposals came down on the side of maintaining economic unity given the fragile position of an expanding, but still weak, economy in Flanders.

On the Walloon side, discussions over greater regional autonomy were much less widespread. The few who were engaged in the debate, however, made proposals as radical as those of their more extreme Flemish counterparts. Meetings of the so-called "Concentration Wallonne," an organization of Walloon activists, during the years leading up to World War II were characterized by a persistent concern with the creation of solidarity and national consciousness among Walloons to counterbalance nationalist and regionalist developments among Flemings.[97] The restructuring of the Belgian state along federal lines was frequently discussed as a way of protecting Walloons and Wallonia from what was seen as a growing threat of Flemish domination. These feelings grew in part out of the gradual but steady displacement of Belgian industry toward the North, the continued growth of Flemish nationalism, and concern over the identification of some Flemish nationalists with the expanding influence of Germany in Europe.[98]

---

[95] Law of June 15, 1935, *Moniteur Belge*, 105, 173 (June 22, 1935), pp. 4002-4016.

[96] Law of July 30, 1938, *Moniteur Belge*, 108, 235 (August 22-23, 1938), pp. 5239-5244.

[97] See generally, Schreurs, *Les congrès du Rassemblement Wallon*, pp. 26-31. See also Jean Duvieusart, *Wallonie 1938* (Liège: Editions de l'Action Wallonne, 1938).

[98] This is not to suggest that fascist ideas were stronger in Flanders than in Wallonia. In fact, the Belgian fascist party Rex got its start in the South. There was little identification of

Most Walloons did not feel any particular identification with the Francophones in Brussels, many of whom were frenchified Flemings. At the same time, many felt strongly that Brussels, as the capital of Belgium, should be a place in which the use of French was in no way restricted. Therefore, in contrast to the Flemish federalist ideas, many of the Walloon proposals were based on the idea of a federalism à trois in which Brussels would be a full constituent unit within the federation.[99] Proposals were also advanced for greater economic integration with France, complete linguistic freedom of choice in Brussels, and the right of all communes along the language border to vote on their choice of regional affiliation.

With growing attention being paid to the restructuring of the Belgian state along language lines, a group of private individuals of various opinions and backgrounds got together to study the issue and report to the government. Their report, which was issued in 1937, rejected federalism on the grounds that it would pose too great a threat to the integrity of the Belgian state.[100] They did, however, advocate that the state be reformed to give recognition to the two "communities" in Belgium. The moderate tone of the report indicates that as World War II approached, Belgians from a wide variety of perspectives were thinking seriously about the language regions as a basis for functional and formal distinctions of a nonlinguistic nature.

World War II and its Immediate Aftermath

Belgium fell quickly to the Germans in the summer of 1940, and many soon came to believe that Germany had won the war. Once again there were a number of collaborators, although both Walloons and Flemings were involved and collaboration was less strongly associated with the Flemish movement than it had been in World War I. Nevertheless, the zeal of some of the right-wing Flemish nationalists again sparked attitudes of resentment. Moreover, just as in World War I, the Germans sought to polarize Belgium along language lines. This was achieved in part by a German decision to liberate most Flemish prisoners of war, while keeping their Walloon counterparts in captivity. In addition, the Germans acted

---

fascism with the Walloon movement, however, whereas Flemish nationalists were seeking to steer foreign policy away from close links with France.

[99] See, for example, Ligue d'Action Wallonne, "Programme de la Ligue d'Action Wallonne," in *Ligue d'Action Wallonne*, collection of pamphlets in the holdings of the Bibliothèque Royale Albert 1er, Brussels, Belgium, 1926-1938, no. 2.

[100] *La réforme de l'état* (Brussels: Centre d'Etudes pour la Réforme de l'Etat, 1937), p. 305.

swiftly to provide complete reparation to collaborators who had been prosecuted after World War I.[101]

The Germans, under orders from Hitler, continued to follow an active *Flamenpolitik* during the war,[102] but its impact was considerably smaller than that of the far-reaching initiatives of World War I. The defeat of the Germans in 1944 led, once again, to a postwar pursuit of collaborators. The activities of some Flemish nationalists during the war and lingering Walloon resentment over the differential treatment of Flemings and Walloons meant that this pursuit took the form of a general suppression of the Flemish cause. The war had served to deepen regional linguistic divisions. In its aftermath, while the Flemish movement was trying to recover from internal schisms and postwar suppression, concern was spreading among a growing number of Walloons over the implications of coexistence with the Flemings in a unitary state in which they were the minority.[103] Feelings surfaced at a meeting of the so-called National Walloon Congress shortly after the war when, in a first *vote de coeur*, delegates voted in favor of union with France, and in a second *vote de raison* they opted for a federal structure in Belgium.[104] Feelings of this sort were not necessarily widespread in Wallonia at the time, but they did represent the continued growth of ethnoregional consciousness among Walloons.

The comments of Fernand Schreurs at the 1945 congress go some way toward explaining the concerns of the delegates. Schreurs pointed to the steady decline in the position of Wallonia in Belgium since 1830, the demographic imbalance between Wallonia and Flanders, the decline in investment in Wallonia, the lack of a significant number of Walloon cultural institutions, the small share of public spending going to Wallonia, and the disproportionate number of Flemings among public servants.[105]

---

[101] Ordinance of September 6, 1940, *Verordnungsblatt des Militärbefehlshabers in Belgien und Nordfrankreich für die Besetzten Gebiete Belgiens und Nordfrankreichs*, 14 (September 10, 1940), pp. 203-205. Borms was appointed to head the commission overseeing the implementation of the law.

[102] See generally, A. De Jonghe, "Het Vraagstuk Brussel in de Duitse Flamenpolitik, 1940-1944," *Taal en Sociale Integratie*, 4 (1981), pp. 405-443.

[103] The 1947 census reveals that there were 4,272,185 people living in the Flemish provinces outside of the district of Brussels as opposed to 2,940,085 living in the Walloon provinces. *Recensement Général de la Population au 31 Décembre 1947*, vol. 1 (Brussels: Ministère des Arraires Economiques, Institut National de Statistique, 1949), p. 172.

[104] *Le congrès de Liège des 20 et 21 octobre 1945: débats et résolutions* (Liège: Editions du Congrès National Wallon, 1945).

These concerns were reiterated at annual meetings during the latter half of the 1940s with appeals to federalist or regionalist ideas increasingly being invoked to confront the growing feeling of *minorization*.[106] What emerges from the discussions at these meetings is a clear sense of ethnic and regional consciousness among the participants. The questions were no longer primarily ones of language. Statistics on public expenditures or the ethnolinguistic background of public officials were based solely on Wallonia-Flanders dichotomies, with no reference to the status of frenchified Flemings, of the Francophones of Brussels, or of intraregional variations. Although attended by only a small minority of Walloons, the conferences were beginning to have some effect in the South, as evidenced by the growing number of publications devoted to Wallonia and Walloon problems in the years following World War II.[107]

Summary

The development of linguistic regionalism among Belgians was a slow process because there was so little to unite either Flemings or Walloons other than linguistic similarities, and even from that standpoint dialectical differences worked against language being a unifying force.[108] The language areas had no tradition of territorial integrity, their economies were closely tied to one another, and power within the Belgian state was highly centralized. The Flemish movement developed as a response to the disadvantaged position of the Dutch language and its speakers in the state. Through the beginning of the twentieth century Flemish leaders were concerned primarily with language rights and Flemish cultural issues. The

---

[105] *Ibid.*, pp. 17-20. In fact, the 1947 census revealed that 39.51 percent of public sector employees came from the Flemish provinces of Antwerp, East Flanders, West Flanders, and Limburg, wheras 31.32 percent came from the Walloon provinces of Hainaut, Liège, Luxembourg, and Namur. Calculated from *Recensement Général de la Population au 31 décembre 1947*, vol.8 (1953), p. 62. The higher administrative and governmental positions, however, continued to be controlled primarily by Francophones.

[106] At the congress in 1949, the delegates adopted a resolution that started with the proclamation that "Wallonia contains today a national minority deprived of all guarantees at the heart of the Belgian state." "Le Congrès de Liège des 1er et 2 octobre 1949," *Nouvelle Revue Wallonne*, 2, 2 (January 1950), p. 9.

[107] For example, Elisée Legros, *A la recherche de nos origines wallonnes* (Liège: Editions des Forces Nouvelles, 1945); *Economie Wallonne*, Rapport Présenté au Gouvernement Belge par le Conseil Economique Wallon le 20 mai 1947 (Brussels: Conseil Economique Wallon, 1947); and the founding in 1949 of the journal *Nouvelle Revue Wallonne*.

[108] A similar point is made by Clough with respect to the development of Flemish national consciousness. *A History of the Flemish Movement in Belgium*, p. 244.

upheavals of war, efforts by Germany to divide Belgium, the extension of the franchise, and sweeping social and economic changes after the war led to the politicization of the Flemish movement, its emergence as a major social and political force in Belgium, and the growth of an organized Walloon response. In the process, the focus of concern in Belgium's political and social life shifted from individual language rights to ethnoregional integrity, and even to territorial autonomy. The growth of regional ideology during this period was not simply a consequence of emerging nationalism but an integral part of evolving intergroup and intragroup relations. Moreover, as regional conceptions became elaborated, accepted, and institutionalized, the nature of Flemish-Walloon relations fundamentally changed.

It is likely that data on economic well-being, religious observance, or modes of livelihood for 1950 would reveal intraregional differences in Flanders and Wallonia that were almost as great as those shown in the figures at the end of the previous chapter. What had changed by the mid-twentieth century was that those differences had become largely overshadowed by the regional dichotomy that had emerged between the North and the South. The linguistic geography of Belgium had become a widely accepted basis for the regional compartmentalization of the state, and this idea had already acquired a degree of formal expression in the language laws of the time. Linguistic regionalism was taking its place as a basic aspect of Belgian political and social life. It was not just an inevitable by-product of ethnolinguistic consciousness but a dynamic component of changing social, cultural, economic, and political arrangements.

Chapter 6

THE RESTRUCTURING OF BELGIUM ALONG LANGUAGE LINES

The development and growth of regional linguistic problems and concerns rendered the unitary structure of Belgium increasingly problematic during the first one hundred twenty years of Belgium's existence as a sovereign state. Matters have come to a head in recent decades, leading to a significant devolution of power along regional linguistic lines.[1] This chapter documents the more important developments in the linguistic regionalization of Belgium which have taken place since 1950. Particular attention is devoted to the role of prior regional ideas and institutions and of more recent social and economic changes in promoting the administrative restructuring of the country along language lines. The implications of recent developments are of such breadth, however, that systematic analysis of them will be deferred to a separate chapter devoted to an examination of the implications of modern regional change in Belgium.

The Growing Salience of Regional Divisions in the 1950s

From the end of World War II through the late 1950s, the Belgian political scene was dominated by two issues: the right of King Leopold III to retain the Belgian throne given allegations of collaboration during the war (the "royal question") and the propriety of government subsidies to private Catholic schools (the "school wars"). Many commentators note that these

---

[1] There is an extensive literature documenting these developments. Among the many works treating the modern restructuring of Belgium, good overviews are: Robert Senelle, *The Reform of the Belgian State*, 3 vols., Memo from Belgium, nos. 315, 319, and 326 (Brussels: Ministry of Foreign Affairs, External Trade and Cooperation in Development, 1978-1980); Theo Luykx and Marc Platel, *Politieke Geschiedenis van België, 2, van 1944 tot 1985* (Antwerp: Kluwer Rechtswetenschappen, 1985); and Kenneth D. McRae, *Conflict and Compromise in Multilingual Societies: Belgium* (Waterloo, Ontario: Wilfrid Laurier University Press, 1986).

matters were of such overwhelming importance that they completely overshadowed the ethnolinguistic issue. While it is true that the 1950s saw comparatively little public debate over language issues per se, regional linguistic divisions had become so important by then that they had a profound impact on the resolution of the royal question and the school wars.[2]

The ethnoregional dimension of the royal question was particularly salient. When Germany invaded and defeated Belgium in World War II, King Leopold III refused to go into exile with much of the rest of the Belgian government. During the occupation he engaged in certain activities that were interpreted by some as making the best of a bad situation, and by others as collaboration.[3] As a result of the taint of collaboration, a significant element of the Belgian population opposed his return to the throne in 1945. Opposition came primarily from the ranks of the Socialist, Communist, and Liberal parties, but many more Walloons were against Leopold's return than Flemings. After prolonged debate, the matter was put to a referendum in 1950. A majority of Belgians (58 percent) voted in favor of the king, but while 72 percent of the voters in Flanders called for his reinstatement, 58 percent of the voters in Wallonia opposed Leopold's return.[4] The ethnoregional division in the vote was accorded considerable attention, and when the government moved to restore Leopold, mass demonstrations broke out in the Walloon industrial towns of Liège and Hainaut. The issue thus served to polarize Flemings and Walloons. With both Leopold and the government recognizing the explosive nature of the situation, the king abdicated in favor of his son Baudouin.

The royal question vividly demonstrated how far ethnoregional divisions had evolved in Belgium. An issue that had initially pitted political parties against one another emerged as a matter perceived primarily in terms of ethnoregional differences. The factors promoting regionalism outlined in chapter 5 intensified in an atmosphere in which attention was focused on a particular sore point for the Flemish movement, collaboration with the Germans. The upshot was that for the first time in Belgian history, a majority in one language region had been able to dictate

---

[2] A similar point is made by James A. Dunn, Jr., "Social Cleavage, Party Systems and Political Integration: A Comparison of the Belgian and Swiss Experiences" (Ph.D. dissertation, University of Pennsylvania, 1970), p. 114.

[3] For a general discussion of the royal question, see E. R. Arango, *Leopold III and the Belgian Royal Question* (Baltimore: Johns Hopkins Press, 1961).

[4] McRae, *Conflict and Compromise in Multilingual Societies: Belgium*, p. 111. In addition, 52 percent of the Brussels population voted against reinstatement.

an outcome contrary to the expressed wishes of a national majority. Moreover, communal and provincial differences were completely overshadowed by the Flanders-Wallonia dichotomy. The fact that the king received a majority of the vote in the two Walloon provinces of Luxembourg and Namur made little difference. The language regions of Belgium were the decisive units.

The question of aid to Catholic schools also took on an ethnoregional character during the 1950s. Not surprisingly, this issue pitted members of the Catholic Party against Socialists, Communists, and Liberals. It dominated the political scene for many years because of the intensity of disagreement and the large percentage of Belgian children attending Catholic schools. Although Belgians remained fundamentally divided along party lines, the greater support for the Catholic party in Flanders and the prominent role played by a Walloon minister in opposing subsidies to Catholic schools led some to frame even this issue in ethnoregional terms.[5] An elaborate compromise was finally reached among party leaders in 1958 that paved the way for renewed attention to specifically ethnolinguistic issues. The significance of these developments is suggested by Lode Claes who, writing in the early 1960s, concluded that "since at least the end of the Second World War it has become general practice in Belgium to interpret every internal political change and event in the light of the bipartite or tripartite character of the country."[6]

Increasing ethnoregional polarization in Belgium was manifesting itself in the 1950s in other ways as well. A group of more radical Walloons organized meetings in 1950, 1953, and 1957 at which regional autonomy was a major focus of discussion.[7] Although attendance was relatively small (fewer than 800 at the 1953 meeting), the publicity surrounding the meetings succeeded in alerting a broader spectrum of the Walloon population to regional issues and provided a unified forum for discussion of Walloon concerns. On the Flemish side, after a period of relative inactivity after the war, a number of influential Flemish nationalists founded the Volksunie in 1954, which continues to be the major political party of the Flemish movement.[8] Members of the Volksunie pressed for federalism, the

---

[5] See generally, Dunn, "Social Cleavage, Party Systems and Political Integration," pp. 53-54.

[6] Lode Claes, "The Process of Federalization in Belgium," *Delta* 6, 4 (Winter 1964-1964), p. 45.

[7] Fernand Schreurs, *Les congrès du Rassemblement Wallon de 1890 à 1959* (Charleroi: Institut Jules Destrée, 1960), pp. 43-49.

[8] The Volksunie celebrated its thirtieth anniversary in 1984 with a volume charting the development of the party and the positions taken by its leaders since 1954. Toon van

establishment of a permanent boundary between Dutch and French-speaking Belgium, and specific limits on the territorial extent of Brussels.

A number of Fleming and Walloon advocates of federalism got together in 1952 to discuss the possibility of a joint federalist proposal. Out of this came the so-called Schreurs-Couvreur bill.[9] The bill called for dual federalism, with Brussels as a federal district. Although some thought the bill signaled a growing Walloon-Fleming consensus on the political structure of Belgium, it failed to address a number of underlying contentious issues, including the exact location of the language border and the expansion of Brussels. Moreover, most Belgians were not at all ready to accept a federal structure, and even many of the Walloons and residents of the capital who were willing to consider the idea thought that Brussels should be a full constituent unit within a federal state. These problems, combined with the preoccupation of the political parties with the school war, meant that the bill failed before it received serious consideration.

In the long-term evolution of the "communities problem" in Belgium, two of the most significant developments after World War II concerned the 1947 census and the activities of a center appointed by the government to investigate solutions to Walloon-Fleming problems. The occupation of Belgium during the war had resulted in the postponement of the 1940 census to 1947. The census had significant implications for the communes along the language border and around Brussels in view of the provisions of the 1930s language laws. If the figures showed a minority of 30 percent or more in a commune, special concessions would have to be made for speakers of the minority language, and a declaration by more than 50 percent of the inhabitants of a commune that their language was different from the regional language would result in a change of regional affiliation. The social and international prestige of French continued to be alluring to some Flemings, and Flemish leaders charged that language questions were being answered on the basis of linguistic aspiration or the desire to change administrative affiliation rather than language practice.[10] The ensuing

---

Overstraeten, *Op de Barrikaden: Het Verhaal van de Vlaamse Natie in Wording* (Tielt: Lannoo, 1984).

[9] See generally, Centre de Recherche et d'Information Socio-Politique, "Tableau synthétique des projets de féderalisme de 1931 à nos jours," *Courrier Hebdomadaire du C.R.I.S.P.*, 129 (November 14, 1961), pp. 3-11; "Autour du manifeste Schreurs-Couvreur," *La Nouvelle Revue Wallonne*, 5, 3 (April 1953), pp. 192-193.

[10] In fact, the director general of the National Institute of Statistics later testified that the language questions on the census had no objective validity. M. Dufrasne, "Communication concernant les recensements linguistiques," Centre de Recherche pour la Solution des Problèmes Sociaux, Politiques et Juridiques en Régions Wallonnes et Flamandes, documents in the collection of the Bibliothèque Royale Albert 1er, Brussels, doc. no. 137.

controversy was so intense that the publication of the census data on language was delayed until 1954.[11]

The census contained potentially disturbing results for both Flemish and Walloon regionalists.[12] The figures indicated that in several communes along the language border and around Brussels, Francophone minorities reached the 30 percent level, leading to charges by Flemings of territorial theft. At the same time, the figures revealed a growth in the population of Flanders by some 385,000 people and a decline in Wallonia of approximately 60,000 people. The Flemings were consequently entitled to more seats in Parliament. The Flemish districts (excluding Brussels) picked up three seats in the House and one seat in the Senate, whereas the Walloon districts lost four House seats and two Senate seats.[13] This change generated considerable concern among many Walloons. The territorial nature of the 1930s laws had rendered the census highly political, and the resulting controversies reinforced regional polarization.

In light of the mounting ethnoregional tensions after World War II, the government sought to diffuse controversy by appointing a commission to investigate the problems of the "Walloon and Flemish regions."[14] The so-called Harmel Center undertook an extensive review of demographic, economic, cultural, and political aspects of the situation during the early 1950s and issued a report in 1958.[15] The Center attracted considerably more attention than had originally been anticipated, and although its report did not seek any immediate legislative changes, it was the major force behind the language laws of the early 1960s.

---

[11] Discussed in Arthur C. Curtis, "New Perspectives on the History of the Language Problem in Belgium" (Ph.D. dissertation, University of Oregon, 1971), p. 371.

[12] *Recensement Général de la Population au 31 Décembre 1947: Repartition au Point de Vue des Langues Nationales Parlées* (Brussels: Ministère des Affairs Economiques et l'Institut National de Statistique, 1954).

[13] The Flemish districts (excluding Brussels) accounted for 104 of the 212 seats in the House of Representatives and 52 of the 106 seats in the Senate in 1958. By 1965 the numbers had increased to 107 in the House and 53 in the Senate. The Walloon districts sent 76 representatives to the House and 38 to the Senate in 1958. By 1965 the numbers had decreased to 72 in the House and 36 in the Senate. Calculated from *Annuaire Statistique de la Belgique*, vol. 80 (Brussels: Ministère de l'Intérieur, 1959), pp. 501, 503; and *Annuaire Statistique de la Belgique*, vol. 87 (Brussels: Ministère de l'Intérieur, 1966), pp. 576, 578.

[14] Law of May 3,1948,*Moniteur Belge-Belgisch Staatsblad*, 118, 156 (June 4,1948),pp. 4560-4561.

[15] Centre de Recherche pour la Solution des Problèmes Sociaux, Politiques et Juridiques des Diverses Régions du Pays, *Rapport Final*, issued as *Documents Parlementaires—Chambre— 1957-1958*, April 24, 1958, doc. 940.

The Harmel Center called essentially for the extension of the territorial principle embodied in the 1930s language laws to bring about a high degree of unilingualism in matters of education and administration. Federalism was rejected, in part because it was not seen as a realistic political alternative at the time. Instead, the Center called for territorial dispersal and decentralization of public services, a definitive determination of the language border,[16] the alteration of administrative and judicial divisions to conform to the language border, the creation of separate Walloon and Flemish cultural councils, the establishment of consultative bodies for Wallonia, Flanders, and Brussels with competence in economic and social matters, and the implementation of strict parity between Flemings and Walloons in central government institutions. For the first time, a government-sanctioned body with representatives from Flanders, Wallonia, and Brussels was calling for a significant alteration of the unitary character of Belgium to reflect ethnoregional divisions.

The Language Laws of the Early 1960s

The 1930s language laws and the subsequent evolution of the Fleming-Walloon question had brought about substantial conceptual and formal distinctions between the language regions of Belgium. Hence, with the resolution of the school wars and the reemergence of ethnolinguistic issues, the debate over possible solutions to Fleming-Walloon tensions centered primarily around regional/territorial concerns. Two developments at the beginning of the 1960s precipitated the enactment of additional territorial language laws. These were the deepening economic crisis in Belgium and the controversy surrounding the 1960 census.

During the 1950s it became increasingly clear that the center of gravity of the Belgian economy was shifting north. The old industrial areas in the South were suffering from decaying infrastructure, the rapid depletion of key natural resources, and a growing inability to attract substantial new investment because of the high cost of labor.[17] Not only was much of the new investment going into the North, but the industrialization of Flanders put heavy demands on the central government. In addition, a generation of young Flemings was emerging that had been educated entirely in Dutch and was therefore in a position to challenge Francophones for positions of responsibility and power.

---

[16] In fact, the report specified the linguistic region to which each of the villages and towns along the language border should belong. *Ibid.*, section politique, p. 73.

[17] For a Walloon perspective on these developments, see Michel Quévit, *Les causes du déclin wallon* (Brussels: Editions Vie Ouvrière, 1978).

Mounting concern among some Walloons over these developments came to a head in the winter of 1960-1961 when domestic economic malaise and the loss of Belgium's important African colony, the Congo (Zaïre), prompted the government to introduce an austerity bill. A flamboyant Walloon Socialist, André Renard, who was the deputy secretary general of the powerful Socialist union, called for a strike to protest the bill.[18] Many workers joined the strike, particularly in Wallonia, but the Christian unions, which were stronger in Flanders, did not participate formally. Even many Flemish socialists were reluctant to go along with Renard, creating a split in the Socialist union along language lines. Although support for the strike did not entirely reflect ethnoregional divisions, it was largely perceived as a Wallonia-versus-Flanders issue. Hugh Clout suggests that the regional split resulted from the "recognition that the north had everything to gain and the south everything to lose from a deterioration of Wallonia's already militant labor relations."[19] While this is undoubtedly true, it is suggestive of a more fundamental point, namely that regional linguistic thinking had become a powerful ideological force in Belgium. The eventual failure of the strike, in part due to the lack of nationwide support, touched off substantial Walloon resentment of the Flemings.

Many commentators date the emergence of Walloon consciousness and mass support for the Walloon cause to the events of the winter of 1960-1961.[20] A significant number of Walloons were thrown together in an atmosphere in which rhetoric focused on the problems of Wallonia and the unwillingness of those in the central government to appreciate and remedy the problems of the South. Moreover, Flemish reluctance to support the strike promoted feelings among many Walloons that advancement of the position of the working class, and of the Walloon economy in general, would only be hindered by the participation of Flemings.[21] Consequently, sentiment spread rapidly in Wallonia for some kind of decentralization along language lines that would give the Walloons greater control over

---

[18] A detailed study of the strike can be found in Valmy Féaux, *Cinq semaines de lutte sociale: la grève de l'hiver 1960-1961* (Brussels: Institut de Sociologie de l'Université Libre de Bruxelles, 1963).

[19] Hugh Clout, *Regional Variations in the European Community* (Cambridge: Cambridge University Press, 1986), p. 85.

[20] See for example, Jacques Lefèvre, "Nationalisme linguistique et identification linguistique: le cas de Belgique," *International Journal of the Sociology of Language*, 20 (1979), pp. 37-58; Albert Henry, "Wallon et Wallonie," in *La Wallonie: Le pays et les hommes*, vol. 1., ed. Rita LeJeune and Jacques Stiennon (Brussels: La Renaissance du Livre, 1977), p. 73.

[21] Curtis, "New Perspectives on the History of the Language Problem in Belgium," p. 499.

their own affairs. Out of these feelings grew the first major Walloon political party, the Mouvement Populaire Wallon, which Renard helped to found. The party was avowedly federalist, and its influence grew significantly in the ensuing years.

In the meantime, the 1960 census was again flaming ethnoregional concerns. Suspicious of the evidence brought forth by the 1947 census of growing frenchification along the language border and around Brussels, many Flemings renewed the charge that language questions were nothing more than referenda over administrative preference, thereby implicitly acknowledging that at least some Flemings continued to be attracted by things French. Walloon leaders denied these claims and further asserted that people should have the right to choose their own language of administration. When the 1960 census forms were distributed, a number of Flemish leaders called on municipal authorities in Flanders not to distribute them.[22] Many complied, forcing the government initially to postpone the census, and eventually to omit language questions altogether.[23] The negative reaction of many Francophones to these tactics served to focus even more attention on the territorial dimension of the so-called "communities problem."

In the aftermath of the strike and the controversy over the census, attention turned to the recommendations of the Harmel Center as a possible way of defusing tensions. The Center had found that the 1930s language laws were not being applied strictly.[24] Flemish concerns over this and the continued influence of French in Flanders, particularly in the area around Brussels, led to two highly emotional marches on Brussels in 1961 and 1962.[25] After the first march, a bill was introduced into Parliament designed to establish the language border in perpetuity.[26] Figure 8 depicts the border that was eventually established. Following the recommendations of the Harmel Center, the bill called for the adjustment of provincial and district lines to coincide with the language border and the elimination of any changes in administrative affiliation. The split over the 1960-1961 strike had

---

[22] A good overview of the controversy surrounding the census is Paul M. G. Levy, *La querelle du recensement* (Brussels: Institut Belge de Science Politique, 1960).

[23] Law of July 24, 1961, *Moniteur Belge—Belgisch Staatsblad*, 131, 182, p. 6140.

[24] E.g., Centre de Recherche pour la Solution des Problèmes Sociaux, Politiques et Juridiques des Diverses Régions du Pays, *Rapport Final*, section culturelle, p. 12.

[25] See, for example, Els Witte et al., coordination scientifique, *Histoire de Flandre des origines à nos jours* (Brussels: La Renaissance du Livre, 1983), p. 328.

[26] *Documents Parlementaires—Chambre—1961-1962*, November 14, 1961, doc. 194-1.

Source: *Atlas of Belgium*. (Brussels: Belgian Information and Documentation Institute, 1985). Plate 2.

Fig. 8. Language Regions of Belgium

made many Walloons more receptive to the idea of a clearly defined boundary between Wallonia and Flanders. The problem was, of course, to reconcile different ideas on the location of the language border.

In some respects this problem was not as difficult as it might seem. Along much of the language border linguistic stability, rather than shift, had been the rule for centuries, and there was little dispute as to the linguistic affiliation of most nearby municipalities. For the most part the language border runs through sparsely inhabited areas with the villages on either side clearly dominated by one language or the other.[27] Problems therefore centered primarily on a few less distinct areas of language change (for example, the Fourons/Voeren communes) and on the communes around Brussels. Many Flemings argued against the further expansion of Brussels or any special concessions to French speakers in the suburbs, exhorting that the "oil stain" of French culture that was spreading out from Brussels into Flanders must be contained. From the perspective of many Walloons and residents of Brussels, however, the Flemings were trying to put a "yoke" on the normal expansion of the metropolitan area, and were sacrificing individual language rights in the process.

Discussion and debate took place for an entire year on the bill. On the Brussels issue, the Flemings rejected a trade-off of greater bilingualism in the capital for an expansion of the capital district, whereas the Walloons were unwilling to accept the denial of all Francophone linguistic rights in the suburbs, particularly those with a substantial French-speaking minority. The other major issue of debate concerned the Fourons communes in the province of Liège and the communes around Comines and Mouscron in southern West Flanders. It was generally acknowledged that the language of most of the area around Comines and Mouscron was French, and the Harmel Center had called for the incorporation of this area into Hainaut. The Fourons communes posed a much more difficult problem. The vernacular of the area is a dialect between Dutch and German, but the communes had long been linked with Liège economically, culturally, and administratively. French was the language of business and government and most children were educated in French beyond the first few years of primary school.[28] Hence, although their mother tongue was of Germanic origin, many of the inhabitants did not feel any identification with Flanders.

---

[27] Levy, *La querelle du recensement*, pp. 75-76.

[28] See generally, Michel Hermans and Pierre Verjans, "Les origines de la querelle fouronaise," *Courrier Hebdomadaire du C.R.I.S.P.*, 1019 (December 2, 1983); "Le problème des Fourons de 1962 à nos jours," *Courrier Hebdomadaire du C.R.I.S.P.*, 859 (November 23, 1979).

The Harmel Center was deadlocked over the Fourons, and therefore did not advocate the transfer of the area to Limburg.

The original bill to establish the language border left both the Comines-Mouscron area and the Fourons communes in their original provinces, presumably to avoid regional exclaves. The manifestly Francophone character of the former area, however, impelled the legislators to call for its transfer to Hainaut. In what was probably as much an exchange of territory as anything else, Flemish representatives insisted on the transfer of the Fourons communes to Limburg. Although there was some controversy in the Comines-Mouscron area over the proposal, dissent was relatively minor. The Fourons population, by contrast, protested vehemently. In a vote organized by Liège provincial authorities, 93 percent opted to remain with Liège.[29] The population of the communes was too small to do much on its own, but the Fourons inevitably became a rallying point and symbol for all of Wallonia as an example of what was often referred to as Flemish imperialism.

Given Flemish dominance in Parliament and mounting public pressure resulting from the second great march on Brussels in October 1962, a version of the original bill was finally passed on November 8, 1962, providing for the transfer of the communes as discussed above.[30] In addition, a number of other less controversial transfers occurred. The law provided for special facilities for speakers of the "other" national language in the Fourons communes, the Comines and Mouscron area, and several other linguistically heterogeneous communes along the language border. The law also modified the 1932 law on the language of administration and education to reflect the new provincial and district boundaries.

The 1962 law was designed to serve as a final determination of the border between Wallonia and Flanders (figure 8), although the Brussels problem was left for later determination. Moreover, the law definitively consecrated the ethnoregional duality of Belgium. By so doing, in the words of one of its most vehement opponents, it sacrificed "the rights of the individual to the rights of soil by promoting a territorial over a personal solution."[31] In fact, as Verdoodt notes, the border was established "without

---

[29] Maurice-Pierre Herremans, "Le bilinguisme et le biculturalisme en Belgique," (Ottawa: Royal Commission on Bilingualism and Biculturalism, 1965), pp. 34-36, cited in McRae, *Conflict and Compromise in Multilingual Societies: Belgium*, p. 153.

[30] Law of November 8, 1962, *Moniteur Belge—Belgisch Staatsblad*, 132, 262 (November 22, 1962), pp. 10315-10319.

[31] Remarks of Representative Janssens, *Annales Parlementaires—Chambre—1961-1962*, January 31, 1962, p. 13. (Author's translation).

serious field investigation and without consulting the population concerned."[32] Given the extent to which tensions between Flemings and Walloons had become territorial in nature, there was arguably no other direction in which to turn, but the law's enactment in no way put an end to territorial disputes.

The Brussels problem was finally addressed the following year in a law concerned with language use in the central government and the capital.[33] Before enactment, this law also went through a long and tortuous consideration process, with amendments being proposed and rejected to annex six new communes to Brussels, and to extend minority language facilities to as many as eight other communes around Brussels, including three in Wallonia.[34] In its final form, the law called for the the division of the old Brussels district into a bilingual district of Brussels-Capital consisting of the nineteen communes that had previously been incorporated into the city (figure 9), a small special district of six communes around Brussels with minority language facilities for the large resident Francophone populations (figure 9), and a unilingual Flemish district of the remaining communes (Halle-Vilvoorde).

Two other contemporaneous laws adapted the educational system and judicial districts to the new territorial arrangements.[35] Together the 1963 laws reinforced the principle of regional unilingualism outside of Brussels, the six peripheral communes, and a few communes along the language border; curtailed the "transmutation" classes described in the previous chapter; introduced strict parity between the language groups in the central government and Brussels; required that official papers and correspondence to individuals or establishments be in the language of the region in which they are found; and, in a particularly controversial provision, stipulated that commercial and industrial entities use the

---

[32] Albert Verdoodt, introduction to the special issue on Belgium, *International Journal of the Sociology of Language*, 15 (1978), pp. 14-15. In view of the controversy over the census, popular consultation or systematic field investigation may not have been politically feasible. Curtis notes that the government insisted on the prerogative of Parliament to alter provincial boundaries under the Belgian Constitution. Curtis, "New Perspectives on the History of the Language Problem in Belgium," p. 534.

[33] Law of August 2, 1963, *Moniteur Belge—Belgisch Staatsblad*, 133, 168 (August 22, 1963), pp. 8217-8233.

[34] *Documents Parlementaires—Chambre—1961-1962*, docs. 331/1-331/26; 331/28-331/34. The Harmel Center had rejected the idea of any communes with special facilities around Brussels.

[35] Law of July 30, 1963, *Moniteur Belge—Belgisch Staatsblad*, 133, 168 (August 22, 1963), pp. 8210-8214; Law of August 9, 1963, *Moniteur Belge—Belgisch Staatsblad*, 133, 161 (August 13, 1963), pp. 8002-8012.

RESTRUCTURING ALONG LANGUAGE LINES

1. Anderlecht
2. Auderghem/Oudergem
3. Berchem-Sainte-Agathe/Sint-Agatha-Berchem
4. Brussels
5. Etterbeek
6. Evere
7. Forest/Vorst
8. Ganshoren
9. Ixelles/Elsene
10. Jette
11. Koekelberg
12. Molenbeek-Saint Jean/Sint-Jans-Molenbeek
13. Saint-Gilles/Sint-Gillis
14. Saint-Josse-ten-Noode/Sint-Joost-ten-Node
15. Schaerbeek/Schaarbeek
16. Uccle/Ukkel
17. Watermael-Boitsfort/Watermaal-Bosvoorde
18. Woluwe-Saint-Lambert/Sint-Lambrechts-Woluwe
19. Woluwe-Saint-Pierre/Sint-Pieters-Woluwe

1. Wemmel
2. Kraainem
3. Wezembeek Oppem
4. Drogenbos
5. Likebeek
6. Sint-Genesius-Rode

Source:
Documents Parlementaires, Chambre des Représentants,
1961-1962, annex, November 14, 1961, no. 194-1.

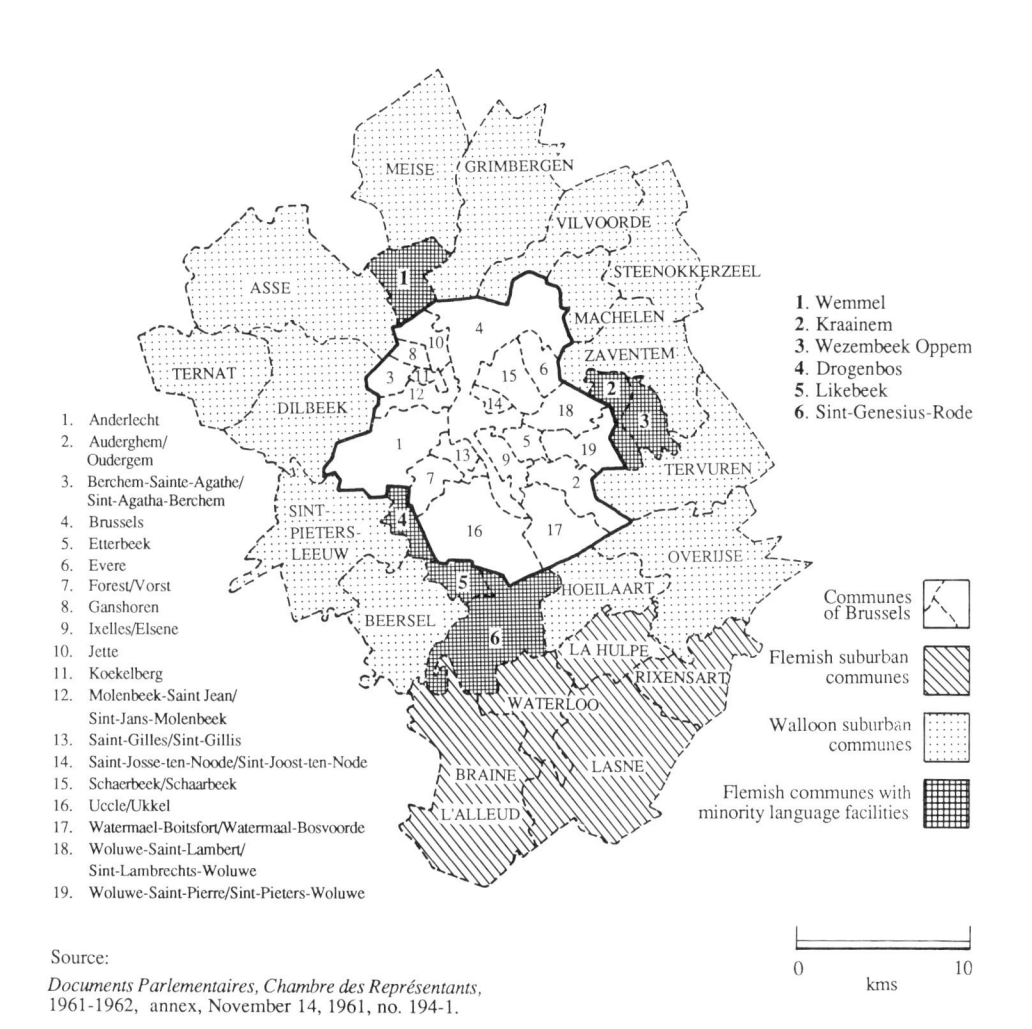

Fig. 9. Communes of the Brussels Metropolitan Area

language of the region in which they are located for acts and documents imposed by law, including those that are directed toward their personnel.[36] Sanctions were included in the laws to discourage violations. Taken together, the 1960s laws were so detailed in their coverage that one commentator concluded that the Parliament had reached the limits of regulatory practicability.[37]

The End of the Unitary State

The language laws succeeded in further solidifying and reifying regional linguistic consciousness. Although the implications of the enactment of these laws will be examined in more detail in the next chapter, mention must be made of a few developments that were of particular importance in bringing about the alteration of Belgium's state structure. First, the consideration of the early 1960s language laws led to a growing consciousness among the population of Brussels of its own particular concerns. There was no particular ethnic basis for this development since many residents of the capital did not feel as if they were either Flemings or Walloons. Many were concerned, however, that the emphasis of both the Flemish and Walloon movements on decentralization worked to their disadvantage. Particular resentment arose over the provisions of the 1960s language laws limiting the growth of Brussels and curtailing freedom of language choice in the suburbs.

As a by-product of the growth of territorial approaches to Fleming-Walloon differences, many people living within Brussels thought of the capital as a region unto itself, with its own particular characteristics and problems. This led to efforts to promote ideas that many regarded as important to the strength and survival of the capital, including freedom of language use and unrestrained growth of the metropolitan area.[38] Support

---

[36] The last provision was especially controversial because it arguably violates article 23 of the Constitution restricting the government's right to pass laws on language to matters relating to the actions of the public sector or to judicial affairs. The rather strained arguments in favor of the provision's constitutionality rested on the idea that article 23 confers an individual right that cannot necessarily be extended to collective undertakings such as business establishments, or that in acting or issuing a document in response to a legal requirement, a private firm becomes a public authority. See Maurice Henrard, *L'emploi de langues dans l'administration et dans les entreprises privées* (loi du 2 août 1963) (Heule: Editions Administratives, 1964).

[37] Curtis, "New Perspectives on the History of the Language Problem in Belgium," p. 558.

[38] See generally, Centre de Recherche et d'Information Socio-Politique, "Le 'Manifeste des 29' et ses répercussions sur les structures politiques de la région bruxelloise I-III," *Courrier Hebdomadaire du C.R.I.S.P.*, 444/445, 448/449, 450 (May 5, 1969, June 20, 1969, June 27, 1969).

came from many Francophones in the surrounding suburbs, as well as a few of the communal councils in the new special district that favored extension of the bilingual regime to their district. Since most of the salient issues concerned matters that would benefit primarily Francophones, it is not surprising that the major organized expression of the Brussels movement was a French-speaking party, founded in 1964 under the name Front Démocratique des Francophones.[39] The party attracted a growing following in the ensuing decade, and succeeded in keeping issues related to the status of Brussels before the electorate.

By firmly establishing the principle of territorial unilingualism in Belgium, the language laws of the 1960s also served to focus attention on areas where the language border did not entirely correspond with linguistic preference. The situation of the Fourons communes has already been described, and this area has continued to be a major source of disturbance. The greatest problem during the 1960s, however, came in the town of Leuven. Leuven is located in the southern part of Flanders and was the site of the Catholic University of Louvain.[40]

After its reorganization in 1833, French became the vehicular language of the university. In 1930 the university began to institute courses in Dutch and eventually Dutch and French programs of study were established. Through the 1960s, however, the French presence at the university remained strong. Moreover, various administrative and educational facilities for French-speaking students, professors, and their families were created in Leuven. With growing competition among the language sections for facilities in the crowded city and rising concerns over territorial unilingualism, Flemish students and political leaders began to call for the transfer of the French part of the university to Wallonia. The matter became a national issue in the late 1960s when the university sought to expand its facilities for French speakers even further. Demonstrations broke out, and the Flemish slogan "Walen buiten" (Walloons out) served to heighten community tensions. Controversy reached such a level that a government eventually fell over the issue and the Francophone section of the university was forced to build a new campus a little south of the language border. The Leuven situation demonstrates the extent to which

---

[39] This is not to suggest that the agendas of Walloons and "Bruxellois" were the same. In fact, some Walloons had come out against the unlimited expansion of Brussels because of their concerns over the concentration of cultural and economic life there. See for example, Fernand Schreurs, "A propos de l'agglomération bruxelloise," *La Nouvelle Revue Wallonne*, 7, 4 (3rd trimester, 1955), p. 238.

[40] The situation is described in some detail in V. Goffart, "La crise de Louvain, du 1er janvier au 31 mars 1968," *Res Publica*, 11 (1969), pp. 31-76.

the linguistic integrity of the language regions had become a major political issue in Belgium. As James Dunn points out: "The spectacle of the attempt to chase the French section from Flanders undoubtedly led to a surge of federalist and separatist sentiment in Wallonia."[41]

The Leuven affair is suggestive of a larger problem. It is clear from the history of the 1960s language legislation that most proponents of the laws saw them as a means of alleviating tensions between Flemings and Walloons. Instead, they arguably served to crystalize regional polarization by firmly institutionalizing the Flanders-Wallonia dichotomy that had developed during the course of the twentieth century. Regional distinctions in turn fostered greater group cohesion defined in territorial terms. The push toward regional autonomy grew out of these developments and was encouraged by a minority complex on the part of both Flemings and Walloons that made significant elements of both groups uncomfortable with the unitary state structure of Belgium.

Flemish minority feelings were the result of a legacy of linguistic disadvantage and economic stagnation. Although post-World War II developments had largely benefited Flemings, minority feelings do not disappear easily. The royal affair in the early 1950s had demonstrated that despite the numerical superiority of the Flemings, as a group they could not necessarily control the direction of future events. Even such a basic recognition of Flemish equality as an official Dutch-language version of the Constitution did not come about until 1967.[42] Several other factors operated to keep minority feelings alive through the 1960s. The top positions in government and business continued to be dominated by a disproportionate number of Francophones, despite the trend toward ethnolinguistic group parity. The sanctions against violations of the 1960s laws did not solve the problem of lax enforcement of the laws, and noncompliance remained a major issue. In addition, a higher proportion of Francophones attended university than did Dutch speakers, and the Dutch-language sector was significantly underrepresented in matters such as funding for scientific research.[43] Moreover, language shift, particularly in and around Brussels, continued to occur, focusing concern on the relative prestige of French and

---

[41] Dunn, "Social Cleavage, Party Systems and Political Integration," p. 161.

[42] Constitutional Revision of April 10, 1967, *Moniteur Belge—Belgisch Staatsblad*, 137, 84 (May 3, 1967), pp. 4771-4777.

[43] See the studies documented in McRae, *Conflict and Compromise in Multilingual Societies: Belgium*, pp. 229-231. McRae's statistics on university attendance reveal that in 1963-1964 53.9 percent of university students were Francophones, as were 51.2 percent in 1968-1969. Moreover, in 1965 the Dutch language sector received only 30 percent of total funding in the natural sciences, social sciences, and humanities.

Dutch. Finally, many Dutch speakers continued to feel uncomfortable using their language in Brussels. A Dutch speaker was not always made to feel welcome in the elegant shops along Avenue Louise or at certain social and cultural gatherings.

Walloons also suffered from distinct feelings of *minorization* in the 1960s. These were fundamentally tied to the economic and demographic decline of Wallonia in the period since World War II. Although the Walloon economy continued to dominate in certain areas, it lost its preeminent position in the decades following the war, with much of the new investment going to the North. Demographically, the Walloon growth rate lagged significantly behind that of Flanders, leading to concern over the numerical superiority of Flemings.[44] This, of course, spilled over into the political arena, with the greater number of Flemish representatives in Parliament translating into a potentially larger say in state affairs.[45] In addition, the increase in government resources going to Flanders in response to economic and population growth in the North fostered concern among Walloons. Perhaps most importantly, the unsuccessful strike of 1960-1961, together with Flemish legislative victories on language issues led to a growing feeling among Walloons that they had little control over their own destinies.

Given this dual minority complex, it is not surprising that pressure mounted for fundamental reform of the unitary state, with a devolution of power along ethnoregional lines. A degree of regional autonomy was an alternative to coexistence in a state in which each group felt dominated by the other. Intense disagreement arose, however, over the extent of autonomy and the status of Brussels within the state. Differences of opinion over this issue reflected ethnolinguistic divisions, with most Walloons and residents of Brussels supporting the idea of a tripartite regional division of Belgium in which Brussels would be one of the regions within the state. The majority of Flemings, by contrast, wanted Brussels to be a *rijksgebied* (capital district) in a state divided into two ethnolinguistic regions. The reasons for this difference of opinion are apparent. As a distinct region

---

[44] See generally, Robert André, "Eléments d'une politique démographique wallonne," *Wallonie*, 74, 5 (1974), pp. 319-323. Between 1930 and 1961, the percentage of the total Belgian population in Wallonia dropped from 37.1 percent to 33.1 percent.

[45] Reference has already been made to changes in representation following the 1960 census (footnote 13). By 1981, the Flemish districts (excluding Brussels) had one more seat in the House of Representatives and four more in the Senate, whereas the Walloon districts had lost two more House seats and retained the same representation in the Senate. Calculated from *Annuaire Statistique de la Belgique*, vol. 87 (Brussels: Ministère de l'Intérieur, 1966), pp. 576, 578; and *Annuaire Statistique de la Belgique*, vol. 101 (Brussels: Ministère de l'Intérieur, 1981), pp. 138, 141.

within Belgium, Brussels would be in a stronger position to control its own affairs, a situation clearly favoring the majority Francophones. Moreover, such an arrangement would potentially pit two regions with French-speaking majorities against one Dutch-speaking region in national matters. As a capital district, by contrast, Brussels would be subject to the dictates of the two regions, with the potential of substantial influence from Flanders, given locational considerations and the numerical superiority of the Flemings in Belgium.

The complex negotiations preceding the revision of the Belgian Constitution in 1970 have been documented elsewhere and need not be repeated here.[46] Suffice it to say that between 1962 and 1970 prolonged and intricate negotiations took place which culminated in a series of substantial amendments to the Constitution significantly altering the unitary character of Belgium.[47] The language laws had in no way changed the centralized political structure of Belgium. Moreover, the laws could always be overturned by a simple majority. Constitutional reform implied a more fundamental recognition of the ethnolinguistic communities through the devolution of at least limited power to the regional level.

The amendments that eventually emerged were the product of elaborate compromises.[48] The following provisions were of particular significance for the regional restructuring of Belgium:

1. The recognition of four linguistic regions (French, Dutch, German, and a bilingual region of Brussels), and three cultural communities (French, Dutch, and German) in Belgium (art. 3 bis and ter);

2. A requirement of parity in the national cabinet and the executive of the Brussels region between members of the French and Dutch communities (art. 86 bis);

---

[46] See, for example, Pierre Wigny, *La troisième révision de la constitution* (Brussels: E. Bruylant, 1972); Paul de Stexhe, *La révision de la constitution belge: 1968-1971* (Namur: Société d'Etudes Morales, Sociales et Juridiques, 1972); and Robert Senelle, *The Revision of the Constitution, 1967-1970*, Memo from Belgium, nos. 132-133 (Brussels: Ministry of Foreign Affairs, External Trade and Cooperation in Development, 1971).

[47] The Belgian Constitution had been revised only twice since 1831, and both revisions involved the extension of the franchise. Although the 1970 revisions dealt with a number of different matters, the most important amendments concerned the "communities problem." Robert Senelle, *The Reform of the Belgian State*, Memo from Belgium no. 179 (Brussels: Ministry of Foreign Affairs, External Trade and Cooperation in Development, 1978), pp. 5-6.

[48] Constitutional Amendments of December 24, 1970, *Moniteur Belge—Belgisch Staatsblad*, 140, 252 (December 31, 1970), pp. 13705-21. Future references to revised articles of the Constitution come from this source.

3. The assignment of all members of Parliament to a Dutch-language or a French-language group (art. 32 bis);

4. The establishment of cultural councils for the French and Dutch communities, consisting of the members of each language group in Parliament, with competence over cultural matters and limited aspects of education (art. 59 bis);

5. The recognition of three regions—the Walloon region, the Flemish region, and the Brussels region—which, by a law passed by a majority of representatives of each language group and two-thirds of the total number of representatives, are to be endowed with an elected body empowered to issue regulations concerning social and economic matters designated by law, with the exception of matters relating to language use or those within the competence of the cultural councils (art. 107 quater);

6. For Brussels, the establishment of a directly elected council divided into French and Dutch-language sections, an executive body that, aside from the chairman, is to consist of an equal number of French and Dutch speakers, and cultural commissions to be elected by each language group with jurisdiction over cultural and educational matters (art. 108 bis and ter);

7. An emergency procedure for both the National Parliament and the Brussels Council through which three-quarters of the members of a language group, by declaring that a particular bill or motion is likely to have a serious effect on relations between the communities, may suspend parliamentary procedure for thirty days while members of the executive branch give a reasoned opinion and invite a vote either on the original bill or motion, or on an amended version (arts. 38 bis & 108 ter); and

8. A requirement of consent by a majority of the members of each linguistic group in the House, as well as two-thirds of all representatives, before the boundaries of the regions can be changed (art. 3 bis).

Other provisions reinforced the linguistic homogeneity of the appellate court districts (art. 104), guaranteed the integrity of the French-language section of the Catholic University at Louvain until its transfer to the Walloon region (art. 132), and allowed for the enactment of laws excepting certain areas from provincial control and submitting them directly to national control (art. 1, a possible solution to the Fourons problem).

The revised Constitution, which charted a path between unitarism and federalism, contained something for everyone.[49] For the unitarists, significant political power was maintained at the center. For Flemings concerned with language shift and cultural domination by Francophone Belgians, cultural issues had been devolved to the representative bodies of the language communities, and the boundaries between language regions had been firmly established with modification requiring Flemish approval. Walloon concerns over diminishing political power and a declining economy were addressed by guarantees of equal representation in the cabinet and the establishment of regional bodies that, although limited in power, provided a potential basis for the regional management of the economy. The Brussels problem was diffused to some degree by a compromise between bipartite and tripartite decentralization in the form of both community and regional institutions with different competences. A number of problems were deferred for later legislative consideration, however, including the status of the Fourons communes and the exact nature and scope of institutions for Brussels and the regions.

Before looking at the subsequent legislation, mention should be made of two laws passed in the same year as the constitutional amendments that were of importance in the evolving regional structure of Belgium. Support from both Flemish and Walloon elements for regionalization of the economy prompted the enactment of a bill creating regional economic planning bodies.[50] The law called for the establishment of economic councils for Wallonia, Flanders, and Brabant (the first time that the names for the language regions appear in a law). The Brabant council was to have an equal number of French and Dutch speakers. The councils were to have consultative and advisory powers and could set up regional development corporations. Although their actions had no binding effect, the establishment of the councils represented an important initial step toward the formal regionalization of Belgium's economy.

The other significant enactment in 1970 paved the way for the constitutional division of Belgium into three regions by eliminating the special district status of the peripheral communes around Brussels.[51] The organization of the communes into a district had been an uneasy

---

[49] See the discussion in J. Brassine, "La réforme de l'état: phase immédiate et phase transitoire," *Courrier Hebdomadaire du C.R.I.S.P.*, 857-858 (October 31, 1979), p. 6.

[50] Law of July 15, 1970, *Moniteur Belge—Belgisch Staatsblad*, 140, 139 (July 21, 1970), pp. 7617-7621.

[51] Law of December 23, 1970, *Moniteur Belge—Belgisch Staatsblad*, 141, 1 (January 1, 1971), p. 17.

compromise from the beginning. Many Francophones assumed this to be a transitional status leading to the complete incorporation of the communes into Brussels. On the other hand many Flemings, thinking of the district as a part of Flanders, vigorously opposed incorporation into Brussels, particularly since that would provide a territorial link between Wallonia and Brussels. After years of serious discord, Parliament finally dissolved the district, incorporated the communes into the Flemish district of Halle-Vilvoorde, and established facilities for Francophones. Primary support came from Flemings, but some Walloons were not anxious to extend the size and influence of the capital either.

The Status of the Language Regions in Belgium Since 1970

The 1970 constitutional amendments provided the basis for the restructuring of the Belgian state along regional lines. The community problem in Belgium had been so strongly cast in regional terms that once the amendments had been enacted, few continued to question whether Belgium's political and economic life should be regionalized along language lines. Instead, debate centered around how this should be achieved and how extensive it should be. Many issues were left unaddressed by the constitutional revisions of 1970, and laws were necessary to implement many aspects of the new constitutional plan. Much of the last decade and a half has been devoted to carrying forward that which was begun in 1970. The process has been slow due to the special majorities required for the establishment of regional representative bodies and the eventual recognition that further constitutional revisions were necessary to expand the basis of regionalization. A brief overview of the most important developments is necessary to an understanding of the current regional structure of Belgium.

The implementation of cultural decentralization proceeded rapidly. A bill was passed on July 31, 1971, dividing Parliament into two language groups and establishing cultural councils for the linguistic communities.[52] This was followed by a law specifying the competence of the cultural councils.[53] In addition, Parliament passed a bill establishing a new level of local government for Brussels and four other metropolitan areas, and enacting a variety of special provisions relating to cultural affairs in the capital, including the reinstatement of the right of the head of the family to

---

[52] Law of July 3, 1971, *Moniteur Belge—Belgisch Staatsblad*, 141, 129 (July 6, 1971), pp. 8449-8453.

[53] Law of July 21, 1971, *Moniteur Belge—Belgisch Staatsblad*, 141, 141 (July 23, 1971), pp. 8910-8912.

decide the language of education for his or her child.⁵⁴ Given the greater strength of the Socialists in Wallonia and the Catholics and Liberals in Flanders, the new cultural councils had a significantly different political composition than did the national Parliament. Hence, after the councils began functioning, the minority political groups pushed for special protections against discrimination along ideological lines. A statute to this effect was adopted in 1974.⁵⁵

The Dutch council was more active than its French counterpart in the early years. Its most dramatic and significant action occurred in 1973 when it adopted a decree calling for the use of Dutch in official communications between all employers and employees in Flanders.⁵⁶ The Francophone press vehemently attacked the decree as unconstitutional and unwise, and passions in Wallonia and Brussels were aroused. Little could be done to change the decree, however, and controversy abated somewhat when sanctions were brought against only a few business establishments, and some of those involved the use of English rather than French.⁵⁷

The French council belatedly responded to the decree with an enactment of its own in 1982 calling for the use of French in establishments in the French-speaking region or in relations with French-speaking workers in general.⁵⁸ The apparent extraterritorial effect of the latter provision conflicts directly with the Dutch decree, but the Dutch council could do little more than protest. Enforcement of the French decree in Flanders was, of course, impossible. Finally, in January 1986 a court established to deal with conflict between councils exempted the communes with minority language facilities from the application of the Dutch decree and limited the reach of the French decree to employers located or employing personnel in the French-language region.⁵⁹

---

[54] Law of July 26, 1971, *Moniteur Belge—Belgisch Staatsblad*, 141, 163 (August 24, 1971), pp. 9782-9808.

[55] Law of July 16, 1973, *Moniteur Belge—Belgisch Staatsblad*, 143, 200 (October 16, 1973), pp. 11706-11710. For a detailed discussion of this law see Christian Daubie, "Le pacte culturel: de sa genèse à son application," *Res Publica*, 17, 2 (1975), pp. 171-200.

[56] Decree of July 19, 1973, *Moniteur Belge—Belgisch Staatsblad*, 143, 172 (September 6, 1973), pp. 10089-10092.

[57] McRae, *Conflict and Compromise in Multilingual Societies: Belgium*, pp. 265-266.

[58] Decree of July 30, 1982, *Moniteur Belge—Belgisch Staatsblad*, 152, 165 (August 27, 1982), p. 9863.

[59] Decision nos. 8 & 24 of the Court of Arbitration on January 30, 1986, *Moniteur Belge—Belgisch Staatsblad*, 156, 30 (February 12, 1986), pp. 1702-1720.

In contrast to the relatively rapid establishment of the cultural councils, the implementation of regional decentralization proved to be considerably more difficult. The necessary special majorities were not easily marshaled. In addition, the issue of economic decentralization was complicated by the prior existence of regional economic councils and development corporations that did not have the same territorial competence as the regional bodies established in 1970. In the absence of special majorities, a first small step toward regionalization was taken in 1974 in the form of a "preparatory" law creating consultative councils consisting of sitting senators from each of the regions.[60] Given the shaky mandate of the councils and their lack of autonomy, however, they were largely ineffective, and were abolished after only three years, leaving the executives as the only regional authority.[61]

Six years after the constitutional amendments of 1970, regional decentralization was far from complete, the institutions that had been established were not well coordinated and the scope of competence of the various bodies was not entirely clear. In the summer of 1976, the political parties therefore selected a group of thirty-six delegates to engage in comprehensive negotiations over regionalization. For an entire year, interrupted by the fall of a government, the representatives deliberated and, despite serious differences of opinion, reached an agreement in May 1977 that was set forth in a document known as the "Egmont Pact" after the palace in which much of the negotiation took place.[62] Although the Pact was never fully implemented, it was very influential in shaping the direction of future reforms. Briefly, it set forth detailed plans for the establishment of powerful regional institutions that, among other things, were to have the power to tax and to enter into certain international agreements. In an ingeneous proposal for the Brussels area, the Pact recommended that residents of the six peripheral communes with facilities

---

[60] Law of August 1, 1974, *Moniteur Belge—Belgisch Staatsblad*, 144, 161 (August 22, 1974), pp. 10368-10371. For an in-depth examination of this law, see J. Brassine, "La régionalisation: la loi du 1er août 1974 et sa mise en oeuvre, I-II," *Courrier Hebdomadaire du C.R.I.S.P.*, 665; 667/668 (December 12, 1974; January 10, 1975).

[61] Law of July 19, 1977, *Moniteur Belge—Belgisch Staatsblad*, 147, 144 (July 27, 1977), pp. 9612-9613.

[62] For a detailed account of the negotiations see Centre de Recherche et d'Information Socio-Politique, "Du dialogue communautaire de l'hiver 1976-1977 au pacte communautaire de mai 1977, I-III," *Courrier Hebdomadaire du C.R.I.S.P.*, 767; 772; 783/784 (June 6, 1977; September 9, 1977; December 16, 1977). The major provisions of the Egmont Pact are set forth in English in *Documents Illustrating the History of Belgium*, vol. 2, *Modern Belgium from 1830 up to the Present Day*, Memo from Belgium no. 180 (Brussels: Ministry of Foreign Affairs, External Trade and Cooperation in Development, 1978), pp. 64-72.

be allowed to elect a fictitious domicile in Brussels proper for electoral, judicial, administrative, and educational purposes.

The ambitious proposals of the Egmont Pact faced legal and political hurdles and, consequently, were considerably altered before enactment. On the legal side, additional constitutional amendments were required to implement certain proposals, and Belgium once again plunged into the complex process of constitutional reform. From a political standpoint, intense opposition surfaced on the Flemish side to the plan for the Brussels area, and the Catholic party, which had a majority at the national level, expressed concern over devolving significant power to the regional level because Catholics would be in the minority in Wallonia. A period of complex negotiation and upheaval followed.[63] The first significant agreement came in July 1979 in the form of a law extending the competence of the community councils to matters such as public health and scientific research, and establishing the power of the regional representative bodies over such matters as employment, housing, water, and energy.[64] After the fall of two governments, a tripartite coalition of Catholics, Socialists, and Liberals was finally able to proceed with constitutional reform by agreeing to defer the Brussels question until a later date.

The new constitutional amendments were passed in the summer of 1980 along with two laws on institutional reform. The constitutional revisions brought about the following changes: (1) converting the cultural councils to community councils with expanded competence and independent executives (art. 3 ter, 59 bis), (2) allowing for enactments of the regional representative bodies to have the status of laws (art. 26 bis), (3) opening the possibility for the fusion of the community and regional councils (art. 59 bis), and (4) establishing a court of arbitration to oversee conflicts between enactments of the various bodies (art. 107 ter).[65] A law adopted by the requisite special majority shortly after the constitutional amendments established the basis for the present regional structure of Belgium.[66]

---

[63] An overview of the major developments during this period can be found in Xavier Mabille, "Les facteurs d'instabilité gouvernementale: décembre 1978—avril 1981," *Courrier Hebdomadaire du C.R.I.S.P.*, 916 (April 10, 1981).

[64] Law of July 5, 1979, *Moniteur Belge—Belgisch Staatsblad*, 149, 131 (July 10, 1979), pp. 7779-7781. In addition, a number of royal decrees were issued in 1979 creating and defining the competencies of community and regional executives. *Ibid.*, p. 8.

[65] Constitutional amendments of July 17 and July 29, 1980, *Moniteur Belge—Belgisch Staatsblad*, 137 and 146 (July 18 and July 30, 1980), pp. 8643-8644 and 8899-8903.

The Constitution gave the community councils authority over cultural affairs and so-called *personnalisable/persoonsgebonden* matters (for example, matters relating to individual needs). The scope of these domains was specified, with cultural affairs to include matters such as education, broadcasting, and tourism, and personal matters to encompass health care, social welfare services, and scientific research (arts. 4 & 5). The regional institutions were endowed with broad powers, including competence over urban planning, environmental affairs, housing, water, regional economic development, energy, employment, and the supervision of provincial and local authorities (arts. 6 & 7). In view of the substantial overlap in the representational basis of the Flemish community council and the regional authority, Flemish legislators incorporated a provision in the law fusing the two bodies (art. 1). The Walloons have not followed suit because of the large number of Brussels residents on the community council and the substantial differences between their objectives and those of many Walloons. As a result, the hierarchy of institutions in Belgium is asymmetrical. Nevertheless, a movement has begun to encourage the fusion of the Walloon regional and community councils for organizational and political reasons.[67]

The initial structure and organization of the government under the special law (art. 28) is represented diagramatically in figure 10. All of the representatives in the National Parliament are assigned to either a French or a Dutch-language group. The community councils are made up of the members of each group.[68] The Dutch community council functions also as the regional council for Flanders, except that the six members from Brussels do not participate. On the French side, the Walloon members of the community council meet as a separate regional council. A regional council for Brussels is also envisioned by the Constitution, but no agreement has yet been reached on its form or competence, and it has not yet been established. Each of the legislative organs also has an executive whose members are divided proportionately between representatives of the unilingual language regions and Brussels. The law envisions eventual reform of the Senate (requiring additional constitutional amendments), after which the councils are to be composed only of directly elected senators, thereby providing a greater degree of distinction between regional/community representational bodies and national legislative institutions.

---

[66] Law of August 8, 1980, *Moniteur Belge—Belgisch Staatsblad*, 150, 158 (August 15, 1980), pp. 9434-9451.

[67] "Premier pas vers une fusion entre communauté et région," *Le Soir*, July 3, 1986.

[68] There is also a community council for the German speakers that will not be dealt with here.

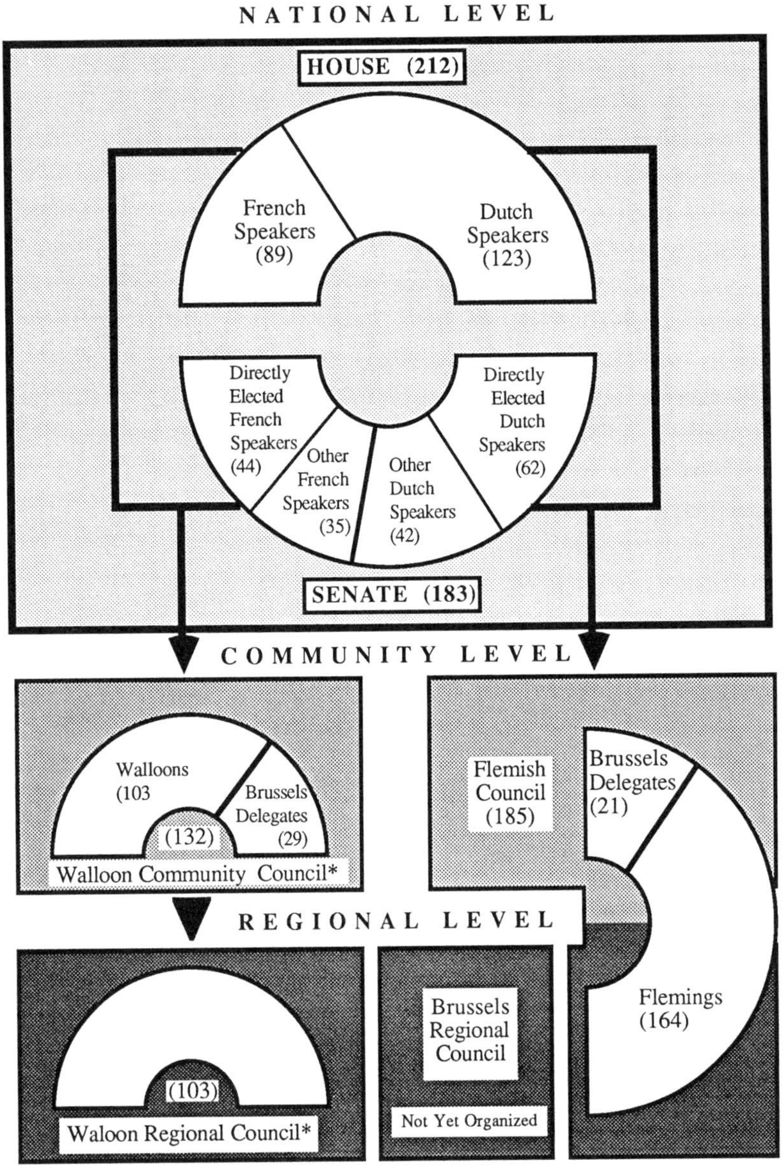

NATIONAL LEVEL

HOUSE (212)
- French Speakers (89)
- Dutch Speakers (123)

SENATE (183)
- Directly Elected French Speakers (44)
- Other French Speakers (35)
- Other Dutch Speakers (42)
- Directly Elected Dutch Speakers (62)

COMMUNITY LEVEL

Walloon Community Council* (132)
- Walloons (103)
- Brussels Delegates (29)

Flemish Council (185)
- Brussels Delegates (21)

REGIONAL LEVEL

Waloon Regional Council* (103)

Brussels Regional Council — Not Yet Organized

Flemings (164)

*The Walloon Community and Regional Councils should have one more member, but a proportional representation provision of the electoral laws of Brabant led to the assignment of one Flemish senator to Walloon Brabant. As a member of the Volksunie, he has not been allowed to take his seat, a matter that is presently before the courts.

Fig. 10. Legislative institutions in Belgium

A law passed at the same time as the special law spells out in detail such matters as the financing of the institutions, conflict resolution, and the relation of provincial and municipal institutions to the community and regional councils.[69] The issue of financing is particularly sensitive. Before 1980 the councils were subject entirely to national government appropriations, leading some commentators to question their true autonomy.[70] The 1980 constitutional revision and the August 9 law theoretically give the councils the right to obtain rebates and levies on centrally imposed taxes, to raise revenue from fees, to procure annual grants from the central government, and to borrow funds (Constitution art. 110, law of August 9, 1980, title 1). As McRae points out, however, "the power to levy taxes independently of the central government is rather narrowly circumscribed, and the Ministry of Finance of the central government remains the sole collection agent for taxes. For at least the initial phase of the new institutions, the predominant source of funds appears inescapably to be direct grants from the central government."[71] Hence, the question of autonomy remains pertinent.[72]

Implementation of the institutional reforms began in October 1981, but its progress has been slow. By November 1981 the new institutions were functioning and the regional and community executives had been established as independent entities. The new institutions have been steadily expanding their role, even into the international sphere.[73] The impasse on Brussels, however, has yet to be overcome. The 1980 reforms were achieved only by putting the Brussels problem aside, and, despite occasional outbursts of optimism in the press,[74] an agreement still seems far off. Brussels remains dependent on the central government, a situation that, by its very persistence, may make it less likely that the capital can emerge as a region with the same status as Wallonia and Flanders since the latter already have established and functioning regional structures.[75]

---

[69] Law of August 9, 1980, *Moniteur Belge—Belgisch Staatsblad*, 150, 158 (August 15, 1980), pp. 9451-9462.

[70] See, for example, René Gérain, "La responsabilité politique des ministres devant les conseils culturels," *Res Publica*, 17, 1 (1975), p. 34.

[71] McRae, *Conflict and Compromise in Multilingual Societies: Belgium*, pp. 169-170.

[72] See the discussion in "Régions et communautés: le pouvoir politique délayé dans la subordination financière," *Le Soir*, October 11, 1985.

[73] The new institutions have representatives in UNESCO and the Council of Europe, and the Walloon regional council has entered into formal agreements with its counterpart in Quebec.

[74] For example, "Enfin le printemps pour Bruxelles?" *Le Soir*, April 18, 1986.

Summary

The shift to a territorial approach to ethnolinguistic issues during the first half of the twentieth century has culminated in recent decades in the restructuring of the Belgian state along language lines. What has emerged is a state in which the most important administrative subdivisions reflect the linguistic geography of the country. Although the reforms have been far-reaching, Belgium is not a strictly federal state.[76] This may be due in part to the traditional concern among Belgium's mainstream political elite that federalism would be a major step toward the breakup of the state. In any case, in view of the lack of control of the new regional/community institutions over their own composition, their lack of true fiscal independence, and the overlap in the membership of regional and national legislative bodies, Belgium cannot properly be thought of as a federation. Nevertheless, the regions and communities have acquired substantial powers that can be altered only by special legislative majorities. Hence, the system is decidedly not unitary. It is a hybrid between unitarism and federalism that is sometimes referred to as "communal."[77]

More important than its label, the restructured political system of Belgium represents a sweeping institutionalization of evolved regional concerns and structures that is profoundly altering Belgium's social, political, economic, and cultural character. The focus of this chapter on the changing structure of government institutions has meant that only scant attention could be paid to these changes. Chapter 7 addresses explicitly the implications of regional restructuring in an effort to provide some insight into the significance of the described changes for the future of the Belgian state.

---

[75] A similar point is made in McRae, *Conflict and Compromise in Multilingual Societies: Belgium*, pp. 320-321.

[76] This issue is discussed in Glenn V. Stephenson, "Cultural Regionalism and the Unitary State Idea in Belgium," *The Geographical Review*, 62, 4 (October, 1972), pp. 501-523.

[77] Senelle, *The Reform of the Belgian State*, p. 139.

Chapter 7

THE CHANGING CONTEXT OF GROUP IDENTITY AND INTERACTION

The regional developments documented in the previous chapter are not simply reflections of ethnolinguistic divisions. They are significantly altering the spatial structure of Belgium, with profound implications for group interaction and the future structure of Fleming-Walloon relations. To put individual developments into some sort of larger perspective, changes in the conceptual, functional, and formal regionalization of Belgium must be recognized as integral to the evolving situation. The underlying proposition of this study is that in reformulating the game as a territorial one, the rules are being changed by the structure and logic of the new game. Some understanding of the new game is therefore a prerequisite to an interpretation of individual rule changes.

The major changes in the formal regional structure of Belgium have taken place since 1970, and the process is far from complete. Certainly the extent to which various social, cultural, political, or economic arrangements have been altered as a result of these changes is not yet at all clear. Any overall assessment would therefore be premature and highly speculative. Rather than either focusing on specific outcomes or proposing broad hypotheses about the future, then, this chapter seeks to identify some of the most important ways in which evolving regional arrangements are shaping the nature and context of ethnolinguistic group concerns and interaction in Belgium.

The changes in the conceptual, functional, and formal regionalization of Belgium can be looked at in two ways: as a reconstruction of the stage on which social processes are played out, and as a modification of the spatial context that forms the basis for certain attitudes and understandings. The former is significant insofar as changing regional arrangements affect group interaction patterns, whereas the latter directs attention to the ways in which territorial changes serve to recast group identity and concerns in different terms. Both of these matters will be explored in an effort to demonstrate the pivotal role of regional change in altering the context of group relations and, by extension, the nature of the problem itself.

Changing Regional Arrangements and Interaction Patterns

Two complementary tendencies can be identified in association with the regional restructuring of Belgium. One is the encouragement of greater homogeneity within linguisitically defined regions, and the other is the creation of more pronounced discontinuities between regions. Aspects of these tendencies are explored below along with an assessment of their implications for group interaction and communication.

Linguistic Considerations

A primary objective of much of the legislation related to language since 1923 has been to foster unilingualism within Flanders and Wallonia. Given the lack of census data on language since 1947 and the suspect nature of the figures from the census of that year, it is difficult to assess the degree to which regional unilingualism has been achieved. In the early twentieth century, the use of Dutch in Wallonia was limited to areas close to the language border and a few concentrations of Flemish workers in the industrial cities of the South. In Flanders, by contrast, French had a substantial presence in most of the larger urban centers, particularly in and around Brussels, Antwerp, Leuven, and Ghent.[1] In public life, French was used rather extensively in the North, while Dutch was extremely rare in the South. Hence, it is primarily to Flanders that we must look for evidence of greater linguistic homogeneity.

Leaving aside for the moment the few communes with minority language facilities around Brussels and along the language border, there is little question that legislative changes have substantially eroded the position of French in public affairs in Flanders.[2] The impact of language legislation on such institutions as the courts and municipal government has already been documented. Aside from communities that are heavily dependent upon Francophone tourism (for example, the North Sea beach resorts), French is also now rarely heard in shops and restaurants, and is specifically forbidden in official business relations in Flanders. A number of

---

[1] See the map of languages in *A Manual of Belgium and the Adjoining Territories: Atlas* (London: Naval Staff, Intelligence Division, 1922), pp. 4-5.

[2] Albert Verdoodt, *Les problèmes des groupes linguistiques en Belgique*, Bibliothèque des Cahiers de l'Institut de Linguistique de Louvain (Leuven: Peeters, 1977), p. 57.

Francophones who lived in Ghent before 1965 and habitually spoke French at school and while shopping testified in interviews to a substantial decrease in the number of establishments in which they now felt comfortable using French. Since World War II bilingual street and business signs have also been virtually eliminated from Flanders. Moreover, the decree of the Dutch Cultural Council requiring the use of Dutch in the private sector has brought about a significant decline in the use of French in business establishments in the North.[3] In fact, in an interview conducted in July 1986, personnel at the national commission established to oversee language questions testified to a sharp decrease since October 1985 in the number of complaints related to the application of this law. They attributed the decline to a drop in instances of noncompliance with the decree.[4]

Educational reforms have also had an effect on the linguistic homogeneity of the regions. As Baetens Beardsmore points out, under the new language laws all education in Belgium is essentially monoglot.[5] With the virtual elimination of French-language schools in Flanders, the current generation of children of French-speaking Flemings is, for the most part, being educated entirely in Dutch.[6] Knowledge of French as a second language may also be declining among Flemings due to the shift away from the study of French as the exclusive second language in Dutch-language schools. In both regions the other principal national language is to be taught as the second language unless a group of at least eight students petitions for

---

[3] Henri Hendrickx, "Le Français dans les secteurs financiers et industriels en Flandre," in *La langue française dans les pays du Benelux: besoins et exigences*, ed. Eddy Rosseel (Brussels: Association Internationale pour la Recherche et la Diffusion des Méthodes Audio-Visuelles et -Globales, 1982), pp. 72-74. The author points out, however, that a knowledge of French is still important in some aspects of private business affairs in Flanders because issues relating to exports to Francophone areas frequently have to be handled in French and branches of businesses centered in Brussels or Wallonia frequently use French at the executive level.

[4] The commission was created by the law on the use of languages in administrative affairs of August 2, 1963, *Moniteur Belge—Belgisch Staatsblad*, 133, 168 (August 22, 1963), pp. 8217-8233, art. 53.

[5] Hugo Baetens Beardsmore, "Bilingualism in Belgium," *Journal of Multilingual and Multicultural Development*, 1, 2 (1980), pp. 145-154.

[6] In 1983-1984, the total number of children recorded as Francophone in the Dutch-language region at any level of schooling (presumably on the basis of attendance of French-language classes) was 3340, or less than 0.3 percent of the student population of Flanders. *Annuaire de Statistiques Régionales—1984* (Brussels: Institut National de Statistique, 1984), p. 69. By contrast, the figures compiled by Clough for 1902 indicate that 2.6 percent of the elementary schools in Flanders were French. Shepard Clough, *A History of the Flemish Movement in Belgium: A Study in Nationalism* (New York: Richard R. Smith, 1930), p. 155. Given the growing importance of Dutch in public affairs and business, few of the Francophone families of Flanders now resist education in Dutch. In fact, many regard it as a substantial advantage, ensuring that their children will be completely bilingual.

a different language. For some time many Walloons have opted for English or German rather than Dutch.[7] French continues to be the overwhelming choice in Flanders, but its position has declined somewhat with the growth in interest in the study of English.[8]

The question of whether these changes are prompting a decline in individual use and understanding of French in Flanders is more difficult to assess. There certainly are far fewer unilingual Francophones living in Flanders than previously, but there is little direct evidence of actual loss of French. With the children of French-speaking Flemings being educated in Dutch, however, many of the younger generation are now using the regional language among themselves. This, combined with the steady progress of Dutch in most aspects of Flemish life, may lead to a gradual decline of French. Moreover, the regionalization of many public sector jobs may have a negative impact on bilingualism as fewer positions require a knowledge of the other major national language.

As for competency in French among the general Flemish population, some facility with the language remains high due to its continued study in school and the legacy of French prestige. Formal language study does not necessarily translate into substantial linguistic ability, however, particularly under circumstances in which there is little opportunity to use the language, as is now the case with French in many parts of Flanders. Although this has unquestionably benefited the status and use of Dutch, it has probably had a negative impact on bilingualism.

In Wallonia, Dutch has never played a significant role, and the recent reforms ensure that it is unlikely to do so. Given the growing importance of Flanders from an economic standpoint and the use of Dutch in many matters that are not strictly Walloon, there is a growing emphasis on the study of Dutch as a second language in the South. Nevertheless, French continues to dominate all aspects of life in Wallonia, and the recent reforms have served to reinforce that reality.

As for the communes without legally mandated unilingualism— Brussels, and the communes around the capital and along the language border with protected linguistic minorities—they have a distinct linguistic character of their own precisely because of their exceptional status. In Brussels, the legislative changes of the past twenty-five years have generally

---

[7] Josiane Hamers, "The Language Question in Belgium," *Language and Society*, 5 (Spring/Summer 1981), pp. 17-20.

[8] Eddy Rosseel, "Le français dans l'education en Flandre," in *La langue française dans les pays du Benelux: Besoins et exigences*, ed. Eddy Rosseel, pp. 76-89. There have been movements in both Flanders and Wallonia to elevate English to the same status as the "other" principal national language.

served to halt the decline of Dutch and to encourage greater linguistic equality between French and Dutch.[9] As noted in the previous chapter, businesses that only twenty years ago were operated primarily in French now generally serve their customers in both languages.[10] In fact, the growing importance of a knowledge of Dutch has prompted some Francophones to send their children to Dutch-language schools.[11] As a result, a struggle to attract students has developed between French and Dutch-language educational institutions. Developments such as this attest to the special linguistic character of Brussels in contrast to Flanders and Wallonia, even though there is no question that French remains the dominant language.

Most of the communes along the language border have long been relatively homogeneous from a linguistic standpoint. In a study based on the results of the 1930 census, Levy found that in 130 of the 177 communes bordering on the legally demarcated boundary between Flanders and Wallonia, more than 90 percent of the population spoke the same language, and in only twelve did more than 30 percent speak a different language.[12] Consequently, only a relatively small number of communes present a problem of regional linguistic affiliation. In seven of these communes, and in another six around Brussels, special language facilities have been established for minority French or Dutch speakers. The opportunities of linguistic minorities in these communes to use their own language in certain matters is an exception to the otherwise strictly sanctioned territorial unilingualism outside of Brussels. These communes present only a spatially and numerically minor exception to the distinct regional language patterns described above, but their political and social importance far exceeds their size.

---

[9] See Freddy Louckx, "Het Taalkundig Integratie-Proces van de Nederlandstaligen te Brussel," *Taal en Sociale Integratie*, 1 (1978), pp. 199-228.

[10] There is little hesitancy now to demonstrate dissatisfaction with the linguistic capabilities of business employees dealing with the public. The author witnessed a shouting match in a supermarket in Brussels in February 1986 after a check-out person replied in French to a Dutch-speaking customer.

[11] Gerda Gielen and Freddy Louckx, "Sociologisch Onderzoek naar de Herkomst het Taalgedrag en het Schoolkeuzegedrag van Ouders met Kinderen in het Nederlandstalig Basisonderwijs van de Brussels Agglomeratie," *Taal en Sociale Integratie*, 7 (1984), pp. 161-208. See also "Quand l'école flamande fait du charme aux francophones," *Le Soir*, April 25, 1984.

[12] Paul M. G. Levy, *La querelle du recensement* (Brussels: Institut Belge de Science Politique, 1960), pp. 75-76.

Demographic Factors

The lack of census data on language makes it very difficult to assess the extent to which the regional restructuring of Belgium has affected where French and Dutch speakers live. Although interregional migration figures are available, they are of little help because they provide no information on the language of migrants. Moreover, it is impossible from such data to ascertain the motives for migration. Hence, we must rely on indirect evidence to obtain some idea of demographic shifts in response to the regional restructuring of Belgium.

At the outset it should be noted that there is no evidence to suggest a mass exodus of French speakers out of Flanders in the last three decades. Moreover, many of those who have moved have undoubtedly done so for economic reasons or convenience. There is, on the other hand, some indication that there has been limited redistribution of the Belgian population along regional linguistic lines. This is suggested by a statistical analysis of listings in the two Belgian "Who's Who" volumes, *Qui est Qui en Belgique Francophone* and *Wie is Wie in Vlaanderen*.[13] The people listed in these volumes are, of course, in no way representative of the Belgian population as a whole. As successful, visible members of the French and Dutch-speaking communities, respectively, however, a comparison of place of birth (invariably before 1960) with present residence provides some interesting data on the demographic characteristics of a segment of Belgian society during the period of regional restructuring. Despite the unrepresentative character of this segment, it is useful to examine because it encompasses people who are likely to have the financial means and information to move in response to something other than basic economic necessity. While many in this category have certainly moved largely for convenience or job opportunity, those factors themselves are likely to have been influenced, at least in part, by the regional restructuring of the country. The patterns outlined below, however, are necessarily only suggestive.

The analysis involved the random selection of four letters with a similar number of entries in each volume. For each of the selected letters, A, K, N, and T, the entries were categorized in terms of where the person was born and where he or she now resides.[14] People born or now residing

---

[13] Rik Dekan, *Qui est qui en Belgique francophone* (Brussels: Editions BRD, 1981) and Rik Dekan, *Wie is Wie in Vlaanderen 1980* (Brussels: Groep Cegos Makrotest, 1980). Although the Dutch language volume is not, by its terms, limited strictly to Dutch speakers, the number of Francophones listed is negligible.

outside of Belgium were not considered. This yielded a total sample of 694 entries from the French-language volume and 706 from its Dutch counterpart. Tables 9 and 10 contain a cross-tabulation of the resulting data.

The tables point to a significantly wider distribution of Francophones than of Dutch speakers in Belgium both in region of birth and present residence. Thus, although 57.8 percent of the Francophones in the sample were born in Wallonia, 26.4 percent were born in Brussels, and 15.8 percent in Flanders; and although 55.9 percent presently live in Wallonia, 36.2 percent reside in Brussels, and 7.0 percent in Flanders. In contrast, the overwhelming proportion of Dutch speakers were born in Flanders (95.2 percent) or currently reside there (93.1 percent), with most both born and residing in Flanders (89.1 percent). Obviously, a high proportion of both Francophones and Dutch speakers live in the region of their birth.

Francophones were also more inclined to migrate than Dutch speakers. The biggest movement in Belgium was of Francophones from Flanders to Brussels. Differentials in place of birth and residence suggest only a small net migration of Dutch speakers from Flanders to Brussels (3.8 percent born but 5.1 percent residing in Brussels, or a net apparent movement of 1.3 percent). There has been a much stronger net movement of Francophones to Brussels (9.8 percent of all Francophones), coming predominantly from Flanders (7.9 percent—the difference between the percentage born in Flanders, 15.8 percent, and the percentage now residing there, 7.9 percent), but also from Wallonia (1.9 percent).

The data generally support the hypothesis of some redistribution of the Belgian population along regional linguistic lines over the past twenty-five years. The resulting patterns certainly suggest that the language regions now play a role in affecting population movements. This cannot, of course, be automatically extended to the population at large or necessarily attributed to the regional restructuring of Belgium. When the data are considered alongside indirect evidence from other sources of demographic responses to the imposition of strict regional unilingualism, the case for an impact of the language laws on population movement becomes stronger. For example, data on language declarations for military service indicate that following the legislative changes of 1963 the percentage of residents of Flanders declaring French as their primary language dropped from 4 percent to 2.6 percent, and by 1977 the figure had fallen to 1.1 percent.[15] Moreover, studies of migration

---

[14] For entries listing two residences, only the first residence was considered. For the occasional entry that listed only a business address, that was used.

[15] Kenneth D. McRae, *Conflict and Compromise in Multilingual Societies: Belgium* (Waterloo, Ontario: Wilfrid Laurier University Press), pp. 279-280.

Table 9. Migration of Francophoness Listed in
Qui est qui en Belgique francophone

N = 694	Region of Present Residence			
Region of Birth	Wallonia	Brussels	Flanders	Total
Wallonia	313 (45.1%)	79 (11.4%)	9 (1.3%)	401 (57.8%)
Brussels	43 (6.2%)	122 (17.6%)	18 (2.6%)	183 (26.4%)
Flanders	32 (4.6%)	50 (7.2%)	28 (4.0%)	110 (15.8%)
Total	388 (55.9%)	251 (36.2%)	55 (7.9%)	694 (100.0%)

SOURCE: Dekan, *Qui est qui en Belgique francophone*, letters A, K, N, and T.

Table 10. Migration of Dutch Speakers Listed in Wie is Wie in Vlaanderen

N = 706	Region of Present Residence			
Region of Birth	Wallonia	Brussels	Flanders	Total
Wallonia	0 (0.0%)	0 (0.0%)	7 (1.0%)	7 (1.0%)
Brussels	3 (0.4%)	3 (0.4%)	21 (3.0%)	27 (3.8%)
Flanders	7 (1.0%)	36 (5.1%)	629 (89.1%)	672 (95.2%)
Total	10 (1.4%)	39 (5.5%)	657 (93.1%)	706 (100.0%)

SOURCE: Dekan, *Wie is Wie in Vlaanderen*, letters A, K, N, and T.

to and from several Walloon cities between 1961 and 1970 show a clear influence of the language border on zones of population exchange, particularly for Charleroi and Namur, with the vast majority of migrants coming from or going to Walloon, as opposed to Flemish, communes.[16]

---

[16] A. Laurent and A. Declercq-Tijgat, "Les migrations internes définitives relatives aux agglomérations de Bruxelles, Liège, Charleroi, Verviers et Namur," *Population et Famille*, 45, 3 (1978), pp. 73-132.

Population movements in the Brussels suburban area also reflect the impact of the territorial language laws. As noted previously, six communes around Brussels have special language facilities for Francophones. Although French speakers have continued to move into all of the communes around the capital, those with language facilities (see figure 9 in chapter 6) have received a disproportionate share of the immigrants. This is borne out in a 1969 survey by Kluft and Jaspers on language use in the communes surrounding Brussels.[17] The study reveals that an average of 48 percent of the residents of the communes with language facilities are Francophones, whereas an average of only 18.4 percent of the residents of the other Flemish communes are French speakers. Legislated territorial unilingualism has almost certainly been a factor in creating this difference.

This conclusion is further supported by the work of De Lannoy and Declerck on the suburban commune of Dilbeek, a commune without minority language facilities.[18] They concluded that most Francophones who moved away from the commune did so for linguistic reasons. Other studies further confirm the greater growth of the Francophone population in the suburban communes of Brussels that have minority language facilities as opposed to those that do not.[19] Thus, demographic patterns around greater Brussels, despite their complex character, provide further evidence of the role of legislated territorial unilingualism in promoting intraregional homogeneity and in diminishing contact among ethnolinguistic groups. This state of affairs is further promoted by the well-documented proclivity of those Francophones who do move to the Flemish suburbs to live together in areas where contact with Dutch-speaking Flemings is minimal.[20]

Comparative population growth rates between Flanders and Wallonia have long been a matter of concern in Belgium, a matter that will be taken up later in this chapter. The demographic responses to regional change described above have important implications for the population

---

[17] Reported in Centre de Recherches et d'Information Socio-Politiques, "L'évolution linguisitique et politique du Brabant," *Courrier Hebdomadaire du C.R.I.S.P.*, 466/467 (January 16, 1970).

[18] Walter De Lannoy and Hugo Declerck, "Migraties naar het Brusselse Randgebied: Het Geval Dilbeek," *Taal en Sociale Integratie*, 3 (1981), pp. 269-280.

[19] B. Jouret, *Définition spatiale du phénomène urbain bruxellois* (Brussels: Editions de l'Université de Bruxelles, 1972); Ivo Driesen and Guido Swalens, "Bevolkingsmigraties in Sint-Genesius-Rode (1945-1975)," *Taal en Sociale Integratie*, 1 (1978), pp. 153-198.

[20] See, for example, Herman Bayens, "The Development of the Brussels Agglomeration," *Delta*, 6, 4 (Winter 1963-1964), p. 89; *Brusselse Randgemeenten: Een Onderzoek naar de Residentiële en Taalkundige Ontwikkeling door de Studiegroep "Mens en Ruimte"* (Antwerp: Kultuurraad van Vlaanderen, 1964).

patterns underlying those concerns. The system is currently structured to encourage most Flemings to stay in Flanders and most Walloons in Wallonia. Higher growth rates in one region could translate into greater representation in Parliament, but territorial expansion is unlikely along most of the language border. The only important exception to this is around Brussels, but this is a major exception. In the communes with language facilities around the capital, the Francophones may be in the process of creating a de facto, if not de jure, expansion of French-speaking territory. This is perceived by many Flemings as being particularly ominous in the commune of Sint-Genesius-Rode because Francophone dominance in that commune would create a corridor between Wallonia and Brussels. It is reasonable to assume that developments in suburban Brussels will be at the heart of the communities problem in the years to come.

Urban Issues

The formal ethnoregional division of Belgium is reflected in anomalies in the expansion of Brussels, incipient changes in the urban hierarchy, and the asymmetrical configuration of the hinterlands of certain major cities located near the language border. With regard to the Brussels issue, there is little doubt that the establishment of a fixed "regional" boundary for the capital, which cannot be changed in the absence of special majorities, is a substantial deterrent to the continued expansion of the city. This is the "yoke" that many Francophones complained about when the language laws of the 1960s were enacted. This is not to suggest that the urban area does not extend beyond the formal boundaries in certain places. There are, however, substantial disincentives for a largely Francophone business to establish its base of operations outside of the nineteen official communes of Brussels, particularly in communes without language facilities. This raises another important issue, the irregular patterns of urban expansion around the capital. The tendency of Francophones to settle in the communes with facilities around Brussels has already been discussed. In addition, many Francophones working in Brussels have chosen to settle in the communes immediately south of the language border near Brussels. These settlement patterns, combined with the tendency of Francophone business interests to orient activities toward French-speaking areas, has produced a pattern of urbanism around Brussels that is heavily skewed to the south.[21] This is particularly noticeable when one compares the largely rural character of the landscape near Brussels to the north and east with the heavily built-up areas to the south, excluding protected forest and park land.

---

[21] Jouret, *Définition spatiale du phénomène urbaine bruxellois*, p. 108.

In terms of the urban hierarchy issue, the restructuring of Belgium has meant a substantial devolution of power to the regional level. Although the regional institutions for Flanders are centered in Brussels, those for Wallonia are split between Brussels and Namur.[22] A lively debate is raging among Francophones as to the proper location for the administration of Walloon affairs. In 1985 a group of Walloon intellectuals, officials, and members of the business community drew up a well-publicized petition opposing attempts to move the regional capital to Brussels. The petition stated that locating regional institutions in Brussels would be a denial of the existence of the Walloon region.[23] If the Walloon movement is successful in retaining the regional capital in Namur, it will alter, at least to a small degree, the urban hierarchy with regard to political matters, with potentially important ramifications for the strong centralizing influence of the national capital. Certainly with some 650 public officials now working in Namur, the city is beginning to play a more important role in the hierarchy of Walloon cities.[24]

The functional hinterlands of major urban centers also reflect the effects of formal regionalization along language lines. Studies by a number of Belgian geographers have revealed that with the exception of Brussels, the zones of influence of the larger Belgian cities are markedly altered by the presence of the language border.[25] Although work is at an early stage on this issue, an examination of Liège and its hinterland by Van der Haegen, Pattyn, and Cardyn established that "language factors limit the range of its commerce and service functions towards the North and the East."[26]

Each of these urban issues tends to confirm the trend toward greater intraregional homogenization and dichotomization between regions. The partial insulation of Flanders from the spread of the Brussels region helps to

---

[22] "L'éxécutif wallon et 250 fonctionnaires à Bruxelles, 650 autres à Namur," *Le Soir*, March 7, 1986. Namur was chosen ostensibly for its central location. It was also the capital of Wallonia during the short-lived administrative partitioning imposed by the Germans during World War I.

[23] Unpublished petition presented at a press conference publicized under the title "Appel aux Wallons," Liège, Maison de la Presse, December 9, 1985.

[24] "Namur peut chanter le <<Bia Bouquet>>," *La Libre Belgique*, March 7, 1986.

[25] Herman Van der Haegen, Martine Pattyn, and C. Cardyn, "The Belgian Settlement System," *Acta Geographica Lovaniensia*, 22 (1982), pp. 283-284; José Sporck and Modest Goosens, "Le réseau urbain: les zones d'influence des villes et la hiérarchie urbaine," in *La cité belge d'aujourd'hui: quel devenir?*, Bulletin Trimestriel du Crédit Communal de Belgique, no. 154 (Brussels: Crédit Communal de Belgique, 1985), pp. 192-197; M. Tilsley, "Het Vestigingspatroon van de Financiële Instellingen in België," *Belgische Vereniging voor Aardrijkskundige Studies*, 48, 1 (1979), pp. 255-296.

[26] Van der Haegen, Pattyn, and Cardyn, *"The Belgian Settlement System,"* p. 284.

maintain the ethnolinguistic integrity of the North, encourages urban expansion to the South, and confirms the special regional status of Brussels. With the transfer of certain governmental functions to Namur, a small challenge is mounted to the centralizing influence of Brussels. Finally, the limiting effect of the language border on zones of influence of major cities promotes functional division along ethnoregional lines.

Changing Political and Governmental Arrangements

The restructuring of Belgium has directly altered the nature and composition of political and governmental organizations in such a way as to reinforce ethnoregional divisions. This is most clearly seen in the structure of such ministries as education and cultural affairs, which are now divided along language lines and which conduct most of their business in the regional language, as well as in the creation of new regional and community institutions. The tangible impact of these developments is substantial. In connection with the establishment of the new community and regional councils alone, more than eight thousand public employees have been, or are being, transferred from central government positions.[27] At the national government level, even those ministries that have remained intact (for example, defense and foreign affairs) are subdivided internally along language lines.[28] These developments have created a situation in which separate institutions, or subdivisions thereof, are evolving different, and sometimes contradictory, approaches to political, economic, and social issues. As a result, these arrangements play an important role in fostering interregional differentiation and competition.

The regional restructuring of Belgium, by interposing a level of government between the national and the local, has also tended further to diminish the competence of provincial and communal administrations.[29] Plavsic asserts that communal governmental institutions are in peril of losing all significant authority.[30] While it is probably too early to judge the eventual fate of local government given continued disputes over the competence of regional governmental organs and questions of power sharing with other levels of government, the potential erosion of local government

---

[27] McRae, *Conflict and Compromise in Multilingual Societies*, pp. 171-172.

[28] Baetens Beardsmore, "Linguistic Accommodation in Belgium," p. 11.

[29] For a general discussion of the present role and future prospects for these levels of government, see Rudolf Maes, *La décentralisation territoriale: situation et perspectives*, report to the Ministry of the Interior and the Public Service (Brussels: INBEL, 1985).

[30] Vladimir S. Plavsic, "Les régions, les provinces et les communes en quête d'autonomie," *Res Publica*, 15, 5 (1973), pp. 915-946.

authority could serve further to reinforce dichotomies at the language region level. Awareness of this has led some Belgians who are concerned about regional divisions to press for a reorganization of the system so as to make the provinces the basic governmental unit below the national level.[31]

The implications of changing regional arrangements are further reflected in the character and structure of Belgian political parties. The process of reform was, in part, encouraged by the founding of a number of avowedly regionalist parties during the 1950s and 1960s. The most important of these, the Volksunie, the Rassemblement Wallon, and the Front Démocratique des Francophones, drew their support almost entirely from Flemings, Walloons, and Francophone residents of the greater Brussels metropolitan area, respectively.[32] The processes of reform created periods of substantial governmental instability that helped focus attention on the agendas of the new parties. They achieved significant electoral success in the 1960s and 1970s, and became a force on the Belgian political scene. In fact, all three parties participated in governments during the 1970s. Although these parties have declined in significance over the last decade, they played an important role in the regionalization process.

The growth and success of the regionalist parties, while significant, is perhaps of less long-term import than the splitting of the three traditional political parties (Christian, Liberal, and Socialist) along language lines.[33] The political upheavals of an ethnolinguistic nature accompanying and following the enactment of the early 1960s language laws created increasing tensions between French and Dutch speakers in these parties. The centrifugal effects of these developments became more pronounced as pressure mounted for the parties to take a stand on ethnoregional issues. Separate linguistic wings formed in the Christian party as early as 1965, with the Socialist party following suit in 1966. The crisis at the University of Louvain produced deep schisms in the traditional parties that were reflected in

---

[31] "Emanciper les provinces pour gêner la régionalisation?" *Le Soir*, May 12-13, 1984. There is only a weak movement in this direction, however, and it does not at present seem likely to succeed.

[32] The Volksunie began in 1954 and has been strongly federalist in orientation, with socialist leanings on social, economic, and military issues. The founding of the Rassemblement Wallon in connection with the strikes of 1960-1961 was described in the last chapter. The Front Démocratique des Francophones dates from the early 1960s and supports complete cultural and linguistic freedom in Brussels, incorporation into the capital of surrounding communes containing a high percentage of French speakers, and the maintenance of Brussels' dominant position in Belgium. For a discussion of some of the other smaller regionalist parties see Centre de Recherches et d'Information Socio-Politiques, " Les partis politiques non traditionnels," *Courrier Hebdomadaire du C.R.I.S.P.*, 101 (March 24, 1961).

[33] The events surrounding the splitting of the political parties are summarized in McRae, *Conflict and Compromise in Multilingual Societies*, pp. 142-144.

voting patterns in the national elections of 1968.[34] Shortly thereafter the Christian party split into two separate organizations, with the Liberal party following suit in 1972 as a consequence of internal disputes over the status of Brussels. Through a system of co-representation of the two language groups, the Socialist party maintained its organizational unity somewhat longer, but it too split in 1978.

The division of the traditional political parties brought with it separate institutional structures, congresses, and electoral platforms. Although the branches of the parties still vote together on many national issues, they often divide over regional matters. The regional character of the traditional parties has almost certainly played a role in the decline in support for the regionalist parties, as the latter no longer stand as the primary representatives of specifically regional issues. More importantly, the organization of the traditional political parties now tends to reinforce ethnoregional dichotomization.

Throughout much of Belgian history, the activities of, and patterns of support for, the traditional parties have been regarded as a cleavage cutting across ethnolinguistic divisions.[35] There was, of course, longstanding recognition that the Socialist party was strongest in the old industrial areas of the South and that the Christian party dominated in much of the rural North. Each of the major parties, however, drew support of varying intensity from all over Belgium. Hence, with only a few exceptions, their political activities prior to the 1960s functioned as a unifying force between the language regions. With the division of the parties into separate ethnoregional branches, however, this role of the political parties has been substantially diminished.

The impact of party divisions on ethnoregionalism is suggested by the increasing frequency of legislative votes in recent years following language lines.[36] One of the final, but most important, obstacles to the formation of the 1985 Martens government was disagreement among the various branches of the traditional political parties over the future of Brussels and the scope of competence of community legislative organs.[37] Many of the various interest groups that gravitate around the major political parties

---

[34] Verdoodt, *Les problèmes des groupes linguistiques en Belgique*, p. 139.

[35] See, for example, Paul H. Claeys, "Political Pluralism and Linguistic Cleavage: The Belgian Case," in *Three Faces of Pluralism: Political, Ethnic and Religious*, ed. Stanislaw Ehrlich and Graham Wootton (Westmead: Gower, 1980), pp. 169-189.

[36] See, for example, "Les blocs flamands et francophones s'affrontent à la chambre autour de la périphérie," *Le Soir*, May 25, 1984.

[37] "Martens au finish: Les congrès attendent," continued as "Les deux derniers obstacles avant Martens VI: la question de Bruxelles et le <<59 bis>>," *Le Soir*, November 22, 1985.

have become regionalized as well.[38] Given the important role played by these interest groups in Belgian politics, this represents an additional important reinforcement of political divisions along ethnoregional lines.

Economic Dichotomization

The regional restructuring of Belgium also has encouraged economic dichotomization. Under the July 15, 1970, regionalization law, separate economic councils were established for Wallonia and Flanders, as were regional development societies with jurisdictions corresponding to the regional division.[39] These bodies serve to expand the scope and significance of economic planning at the regional level, a process that began in a climate of growing ethnoregionalism after World War II with the founding of the Conseil Economique Wallon in 1945 and the Economische Raad voor Vlaanderen in 1952. Under the 1980 language laws the powers of regional and community institutions were substantially expanded in the economic sphere, endowing them with primary responsibility for matters ranging from economic policy to land use to the protection of the environment.

These changes, although not yet fully implemented, are beginning to create significant discontinuities in the Belgian economy. In a recent study of economic planning in Belgium, Swyngedouw concluded that the regionalization of some, but not all, economic matters created a complex set of structures that is profoundly impairing the efficiency and feasibility of long-term planning and putting a tremendous burden on multisectoral coordination and integration.[40] In the process, economic regionalization is further solidified and regional differences are accentuated. This is exemplified by changes in the management of water resources. A national water regulatory body was established in 1913. As much of the country's water comes from sources in the South, national management has some distinct advantages. As part of the general regionalization movement, however, the national body was divided into Flemish and Walloon branches.[41] This is creating a

---

[38] John Fitzmaurice, *The Politics of Belgium: Crisis and Compromise in a Plural Society* (London: C. Hurst & Co., 1983), p. 210.

[39] Law of July 15, 1970, *Moniteur Belge—Belgisch Staatsblad*, 140, 139 (July 21, 1970), pp. 7617-7624. A development society was also established for Brussels proper, but the capital is under a regional economic council with jurisdiction throughout Brabant. This is a further example of a compromise that reinforces uncertainty over Brussels' future status as a region and which causes difficulties in planning.

[40] Eric Swyngedouw, *Contradictions between Economic and Physical Planning in Belgium*, working paper no. 3 (Villeneuve d'Ascq: Johns Hopkins European Center for Regional Planning and Research, 1985), section 1.

[41] See generally "Les régions se jettent à l'eau," *La Libre Belgique*, April 23, 1986.

wide range of conflicts between the regions over the control of water pollution and water allocation questions.

The water issue is indicative of the interregional competition that can be generated under present arrangements. Competition is particularly acute over issues of national revenue allocations for economic development projects. The essential features of this problem are outlined below in connection with a separate discussion of interregional competition. It is important to note here, however, that projects funded in Flanders often must be offset by funding for other undertakings in Wallonia, and vice-versa. In the process, economic efficiency can be sacrificed. The major controversies that have surrounded such allocation issues have added to the pressure to devolve economic development spending powers to the regional level.

Since many of the regional economic institutions are only now being organized, it is too early to determine their influence on interregional interaction. The most profound effects appear to be in areas with changed provincial and regional affiliation. In the case of the Fourons, its historical economic links are behind much of the dissatisfaction with the administrative change. There is little evidence yet of substantial alterations in interregional investment patterns or the organization of most businesses following economic regionalization. On the other hand, in the wake of the recent changes, calls have been made for the division of certain state or parastatal organizations along language lines including the national airline, the Office of Foreign Commerce, and the Office of Cooperation in Development.[42] Attention has focused particularly on a few large Belgian banks such as the Caisse Générale d'Epargne and the Crédit Communal de Belgique, although a restructuring of these institutions seems unlikely in the near future.[43] More important in the short term is the reinforcement of the region's separate ethnolinguistic character by the regional language laws, which require that firms conduct business in the language of the region in which they are operating.

Social and Cultural Divisions

Few areas of Belgian life have remained untouched by the process of linguistic regionalization. The division of a wide array of cultural and social institutions along language lines is a case in point. Although there have been distinct Flemish and Walloon cultural institutions since the nineteenth century, their number has greatly expanded in recent years, and many previously unified institutions have divided into separate linguistic

---

[42] McRae, *Conflict and Compromise in Multilingual Societies: Belgium*, p. 198.

[43] "L'Argent flamand dans les mains flamandes," *La Libre Belgique*, March 22-23, 1986.

branches. Included in this category are educational institutions, libraries, broadcasting services, and professional associations. In a regionally divided Belgium, no institution seems immune from pressures to divide. The national library withstood such pressures, although the price was dividing the catalog by language, a major task that has added to the clerical burdens of that financially strapped institution. Even a seemingly neutral organization such as the national opera has been obliged to confront the regionalization issue, although no changes seem likely in the near future.[44]

Those institutions that have divided have had an important impact on Belgian society. Reference has already been made to the departure of the French-language section of the great Catholic University of Louvain in 1968 to the Walloon town of Louvain-la-Neuve. The University of Brussels split as well in 1969, with the Dutch-language section of the university moving to a new campus nearby. In the Brussels case, however, the division was accomplished without significant incident. Another interesting example is that of the Brussels Bar Association, which split into French and Dutch-language sections in early 1985. Although partnership between members of the two bars is permitted and indeed widely practiced, difficulties may lie ahead if the two bars adopt different rules. In the case of the Belgian broadcasting services, the unitary state radio and television concern was restructured in the early 1960s into autonomous linguistic organizations, and in 1977 they were separated entirely, with each division coming under the authority of a distinct community council. As a consequence, the networks have developed in significantly different ways. McRae captured the importance of this in the following statement:

> The new structures provide no explicit mandate for radio or television to promote integrative values, or even mutual understanding, across linguistic boundaries. Instead they have tended to encourage linkages with the Francophone and Netherlandic language communities outside of Belgium, and the current physical setting of the broadcasting system, with its divided headquarters building in Brussels, its separate studios, technical services, canteens—and even separate parking garages—tends to reduce personal contact between the professional staffs ... to a minimum.[45]

Although the formal division of some institutions has been avoided, their internal structures have been drastically altered to reflect ethnolinguistic differences. Such has been the case with the major trade unions. As noted in the previous chapter, the members of the important Socialist trade

---

[44] "En présentant la saison 85-86, Gérard Mortier dit non à la régionalisation de l'Opéra National," *Le Soir*, May 3, 1985.

[45] McRae, *Conflict and Compromise in Multilingual Societies: Belgium*, p. 249.

union divided in 1960 over the general strike. After a period of negotiation, an agreement was reached allowing the union to continue as a single entity. Under the terms of the agreement, however, each linguistic group effectively was given veto power within the organization.[46] More recently the union has gone so far as to establish separate regional groups with substantial autonomy to confront problems at the regional level, and pressure is mounting for an even greater division of power along ethnoregional lines.[47] Divisive internal pressures have caused organizational changes in the other major unions as well. Hence, even though some organizations remain formally united, their internal structure can be a factor in reinforcing regional divisions.

Apart from institutional changes, it is very difficult to generalize about the role of linguistic regionalism in altering patterns of social and cultural interaction and communication. Nevertheless, there are some telling indications that intergroup contact is discouraged under the present regional arrangements. This was suggested in the work of the Belgian geographer Jacques Charlier. In a study of telephone contacts in Belgium in 1982, he showed that if Brussels is not taken into consideration, a pattern of contact emerges that corresponds strongly to the language regions.[48] Analysis of marriage records from communes along the language border also reveals a decrease in interregional marriages over the past twenty-five years, particularly outside of the communes with minority language facilities.[49]

Perhaps the most striking indicator of the impact of altered regional arrangements on interaction and communication is the emergence of an ethnoregional information dichotomy. Belgians are constantly confronted on the one hand with information broken down by language region and, on the other, with much greater material about their own region than about the country as a whole. To provide just a few examples of the former point,

---

[46] Albert Verdoodt, *Linguistic Tensions in Canadian and Belgian Labor Unions* (Quebec: International Center for Research on Bilingualism, 1977), pp. 32-33.

[47] "La F.G.T.B. veut fédéraliser << à l'Envers>>," *Le Soir*, October 15, 1985.

[48] Jacques Charlier, "Les flux téléphoniques interzonaux en Belgique en 1982: une approche multivariée," Paper presented to the Study Group of the International Geographical Union on the Geography of Communication, Montpellier, France, November 18-19, 1985.

[49] This decrease is suggested by the preliminary results from an extensive study currently being conducted by Bernadette Maegerts of the Centrum voor Bedrijfseconomie, Vrije Universiteit te Brussel. The reduced, but still evident, impact of the language border on marriages in a region with language facilities is confirmed by the study of Mullier in the Mouscron area. Jean-Luc Mullier, "La géographie des mariages et les modèles gravitaires: le cas de Mouscron," Mémoire de License en Géographie, Université Catholique de Louvain, 1980-1981.

before 1963 no official statistics were gathered by language region.[50] Now the language regions provide the primary basis for the subdivision of data, with special volumes being published by the National Institute of Statistics devoted to regional matters.[51] In addition, electoral results began to be presented by language region in the early 1960s,[52] many studies have appeared over the past three decades exploring regional or community differences,[53] employment statistics are now usually broken down by language region,[54] and in recent years the media has given substantial play to the debate over comparative demographic trends in Wallonia and Flanders.[55] In fact, virtually no issue passes the notice of the Belgian press without some analysis of regional contrasts. Moreover, the countless surveys that have been taken dealing with issues of regional affiliation serve further to imprint the regional dichotomy in the Belgian conscience by phrasing the questions and presenting the results in ethnoregional terms.[56]

The other side of the information dichotomy is suggested by the trend in the past three decades for Walloons largely to be presented with information about Wallonia and Flemings about Flanders. There has been nothing short of an explosion of literature dealing solely with a single region, usually published in the language of that region. This includes

---

[50] A. Dufrasne, "Au sujet de la communication <<interaction des problèmes linguistique et économiques en Belgique>>," *Journal de la Société de Statistique de Paris*, 104, 7-9 (July-September 1963), pp. 180-181.

[51] *Annuaire de Statistiques Régionales* (Brussels: Institut National de Statistique, 1976-1986).

[52] Centre de Recherche et d'Information Socio-Politique, "Les élections législatives du 26 mars 1961," *Courrier Hebdomadaire du C.R.I.S.P.*, 104 (March 31 and April 7, 1961).

[53] For example, Jan Kerkhofs, "Orientations dans le domaine de l'éthique," in *L'Univers des Belges*, ed. Rudolf Rezsohazy and Jan Kerkhofs (Louvain-la Neuve: CIACO, 1984), pp. 37-67; Henry Zoller, "Les différences objectives entre Flamands et Wallons," Mémoire de Licencié en Sciences Politiques et Sociales, Université Catholique de Louvain, 1963).

[54] R. Leroy, A. Godano, and A. Sonnet, "La configuration spatiale de la crise de l'emploi," *Courrier Hebdomadaire du C.R.I.S.P.*, 1023/1024 (December 23, 1983); Etienne Van Hecke, "Finances et fiscalité communales, analyse cartographique," *Courrier Hebdomadaire du C.R.I.S.P.*, 1017/1018 (November 25, 1983), p. 2.

[55] For example, "La course démographique des provinces belges," *Le Soir*, April 24, 1984, which, despite its title, leads off with a discussion of Wallonia-Flanders differences in growth rates.

[56] See for example, Nicole Delruelle and André-Paul Frognier, "L'opinion publique et les problèmes communautaires," *Courrier Hebdomadaire du C.R.I.S.P.*, 927/928 (July 3, 1981). The scientific value of these surveys has been strongly criticized because the questions themselves suggest to the subjects the importance of the language region as a focus of identification. See Albert Verdoodt, "Les problèmes communautaires belges à la lumière des études d'opinion," *Courrier Hebdomadaire du C.R.I.S.P.*, 742 (November 12, 1976).

atlases, histories, and analyses of political, social, economic and cultural developments.⁵⁷ In the schools Walloon children read of the glories of the great independent prince-bishopric of Liège, while Flemish children are steeped in tales of the grandeur of the great cities of Flanders. In fact, the regional bias of history and geography courses is a widely discussed feature of Belgian education.

The information dichotomy is not just found in books and the classroom. A survey of the two leading liberal Belgian newspapers, *La Libre Belgique* and *De Standaard*, over a two-month period in the spring of 1986 revealed that only 23 percent of the articles in the French-language paper reporting local events in Belgium were based on happenings in Flanders, and less than 7 percent of such articles in the Dutch language paper covered stories from Wallonia.⁵⁸ Even works such as *Who's Who in Belgium* and the *Belgian Historical and Geographical Dictionary of Communes* have been divided so that the volumes in French deal with Wallonia and Brussels, and the Dutch volumes cover Flanders.⁵⁹ Thus, the communal dictionary has been changed from a format in which both the Dutch and French versions covered all Belgian communes to one in which the French volumes list only the Walloon and Brussels communes and the Dutch volumes cover the communes of Flanders and Brussels.

It is clear that greater social and cultural dichotomization along ethnoregional lines has accompanied the regional restructuring of Belgium. The impact this is having on interaction and communication is well illustrated by a recent television series and book produced by the Dutch language radio and television station. The title of the project was *De Andere Belgen*,

---

⁵⁷ A few works representative of this trend are Rita Le Jeune and Jacques Stiennan, eds., *La Wallonie: Le pays et les hommes*, 4 vols. (Brussels: La Renaissance du Livre, 1977-1981); Els Witte, ed., *Geschiedenis van Vlaanderen* (Brussels: La Renaissance du Livre, 1983); *Atlas de la Wallonie*, 5 vols. (Namur: Société de Développement Régional pour la Wallonie, 1980); *Vlaanderen: Een Geografisch Portret* (Antwerp: Gid, 1983); Centre d'Etudes et de Recherches Urbaines, *Rénovation urbaine en Wallonie et à Bruxelles* (Brussels: Fondation Roi Baudouin, 1980); M. Van Naelten, *Systeemtheoretische Verdenning van de Stedelijkheid in Vlaanderen*, 2 vols. (Antwerp: Standaard Wetenschappelijke Uitgeverij, 1961); *Le développement de la Wallonie et les nouvelles technologies* (Liège: Fondation André Renard, 1982); and *De Ontwikkeling van de Vlaamse Economie in Internationaal Perspectief* (Brussels: Gewestelijke Economische Raad voor Vlaanderen, n.d.).

⁵⁸ The survey involved the daily editions of each paper between April 15 and June 15, 1986. Only the major news section of each paper was consulted, and only those articles exceeding ten centimeters in column length were considered.

⁵⁹ Dekan, *Wie is Wie in Vlaanderen* and Dekan, *Qui est qui en Belgique francophone*; Hervé Hasquin, ed., *Communes de Belgique: dictionnaire d'histoire et de géographie administrative. Wallonie*, 2 vols. (Brussels: La Renaissance du Livre, 1980) and Hervé Hasquin, ed., *Gemeenten van België: Geschiedkundig en Administratief Geografisch Woordenboek. Vlaanderen*, 2 vols. (Brussels: La Renaissance du Livre, 1980).

(The Other Belgians) and its stated purpose was to provide information for those living in Flanders about Francophone Belgium.[60] The very existence of such a project is indicative of the limited extent of interregional interaction and communication in contemporary Belgium.

Landscape Factors

As previously noted, the language border runs primarily through sparsely inhabited areas, with most villages to the north having a distinctly Flemish ethnolinguistic character and those to the south being unmistakably Walloon. It traverses an area generally characterized by light road traffic because of the scarcity of industry and the lack of large settlements in the area.[61] Along much of the boundary, there are no dramatic changes in the landscape as one passes from one side to the other, perhaps an indication of the recency of functional and political regional divisions along language lines. In the few cases in which the boundary passes through a more densely populated area (for example the northern part of Engheim) one can observe a different language on street and commercial signs on either side of a road, but the border itself is often not marked.

Certain generalizations can be made about architectural and land use differences between northern and southern Belgium,[62] but there is no striking indication of this at the language border itself. A series of interviews and observations of daily activity patterns of people living in communities adjacent to the boundary did reveal a tendency for people to remain within their own language region. This is not particularly surprising in view of the location of most settlements well within one region or the other. Moreover, few seemed to think of the boundary as a significant obstacle to movement, and bilingualism is more in evidence in the vicinity of the boundary than in other areas outside of greater Brussels. In fact, a relatively recent study of intergroup attitudes revealed that those living close to the language border are more likely to have a positive impression of the members of the other ethnolinguistic community.[63]

---

[60] Guido Fonteyn, Loes Van Mechelen, Eliane Van den Ende, and Herman Van De Vijver, *De Andere Belgen* (Brussels: BRT—Open School, 1984).

[61] Van der Haegen, Pattyn, and Cardyn, "The Belgian Settlement System," p. 315.

[62] Thus, stone is a common building material in the South, whereas many of the older buildings in the North are constructed of wood or brick. Moreover, the North has large expanses of grain agriculture, whereas the terrain of much of the South is better suited to grazing and dairy farming. See generally, Pierre George and Robert Sevrin, *Belgique, Pays-Bas, Luxembourg* (Paris: Presses Universitaires de France, 1967), chs. II-III.

An important exception to these general patterns exists in such disputed border areas as the Fourons and the communes around Brussels. In the latter case, the significant Francophone presence in the communes with language facilities is evidenced by the use of French on a number of commercial signs. Moreover, French can be heard in many commercial establishments. In these areas, and even within the nineteen communes of Brussels, paintbrush battles over bilingual signs are in evidence as zealots from each language group obliterate place names in the "other" language. Landscape indicators of tensions are minor in the Brussels area, however, in comparison with the Fourons. In that district paintbrush battles have escalated to the point where navigation can be exceedingly difficult. Even more striking is the abundance of grafitti that can be seen throughout the district on the pavement, overpasses, walls, and churches, mostly in French, proclaiming "Retour à Liège," "Fourons Wallon," "Vive Liège," "Liège Toujours," "Non au Limbourg," "La Lutte Continue," "Bienvenue en Wallonie," "Non à la Flamandisation," and other less decorous expletives.[64] One is thus made immediately aware that this is an area of considerable dispute.

Away from the language border, the previously noted removal of bilingual and Francophone signs from Flanders has created an almost exclusively Dutch-language visual environment. The movement toward regional unilingualism has prompted fervent responses to cases in which French signs remained visible after the early 1960s. In fact riots broke out over the presence of French signs in certain North Sea resort towns even though they were primarily for the benefit of visiting tourists from France. The efforts of both regions to further strict unilingualism is evidenced by the exclusive use of the French language version of Flemish place names on signs in Wallonia and vice versa. The potential for confusion for the uninitiated is evident, with signs in Wallonia directing motorists to Malines (Mechelen), Anvers (Antwerpen), or Gand (Gent), and their Flemish counterparts referring to Liège as Luik, Mons as Bergen, and Lille as Rijsel. In Brussels, both languages are in evidence, although concern over linguistic parity has reached the point that scholarly articles address questions of the prevalence and positioning of billboards in French and Dutch.[65]

---

[63] Joseph Nuttin, "Het Stereotiep Beeld van Walen, Vlamingen en Brusselaars: Hun Kijk op Zichzelf en op Elkaar," *Mededelingen van de Koninklijke Académie voor Wetenschappen, Letteren en Schone Kunsten*, Klasse der Letteren, 38, 2 (1976).

[64] These and other graffiti were observed during several crossings of the Fourons commune on May 26, 1986.

[65] Stella Tulp, "Reklame en Tweetaligheid een Onderzoek naar de Geografische Verspreiding van Franstalige en Nederlandstalige Affiches in Brussel," *Taal en Sociale Integratie*, 1

Evolving Attitudes and Issues

Each of the issues evaluated above provides some evidence that the reforms of the past few decades, by creating and reifying territorial distinctions of a linguistic nature, have served to reinforce regional divisions and to limit opportunities for interaction between groups.[66] This is necessarily having an important impact on collective identity and mutual understanding, changes that themselves are profoundly affecting the evolution of the regional arrangements and interaction patterns discussed above. At the same time, the very nature of the debate is being reformulated as attention has come to focus on territorial disputes, regional inequalities, and general regional needs rather than purely linguistic issues. A brief evaluation of these matters follows.

The Promotion of Collective Ethnoregional Identity

In a previous chapter it was suggested that the Flemish movement in the early twentieth century adopted a territorial approach in part to encourage the growth of Flemish nationalism. It is hard to refute the success of that strategy. Symbols of Flemish national identity are now widespread in northern Belgium. The anniversary of the Battle of the Golden Spurs is celebrated vigorously as the Flemish national holiday. A huge map of Flanders adorns the great square in the center of Brussels during holiday festivities, the "Vlaamse Leeuw" (the Flemish national song) can be heard often,[67] many publications extol the wonders of Flanders and Flemish nationalism,[68] and the Flemish flag is displayed in the streets on national and regional holidays and is in evidence at many gatherings.[69] Symbols of Walloon identity are somewhat less prevalent, a reflection of a less cohesive sense of community in the South. Nevertheless, a national holiday for the French-speaking community is observed, if not wildly

---

(1978), pp. 261-88. The author found that the number of billboards in Dutch (27.7 percent) was satisfactory given the percentage of Dutch speakers in the capital, but that their positioning gave the impression that Brussels was primarily a Francophone city.

[66] Van der Haegen, Pattyn, and Cardyn describe the effect of the language boundary in the contemporary Belgian administrative structure as a "relation inhibiting factor." "The Belgian Settlement System," p. 345.

[67] *Vlaamsch-Nationalistisch Liederboek* (Vilvoorde: Pieter Céonen, n.d.).

[68] For example, *Vlaanderen ons Vaderland* (Antwerp: De Nederlanden, 1980); and the magazine *Thuis Zijn*.

[69] Some of these symbols are analyzed in Pierre Servais, "Le sentiment national en Flandres et en Wallonie: approche psycholinguistique," *Recherches Sociologiques*, 2 (December 1970), pp. 123-144.

celebrated, magazines such as *Wallonie* perform the role of regional boosterism, and institutions such as the *Musée de la Vie Wallonne* and the *Institut Jules Destrée* seek to document and encourage Walloon political and cultural identity. Even a weak sense of Brussels identity can be said to exist, although it is often in reaction to Flemish and Walloon nationalism and has little symbolic expression.

This is not to suggest that ethnoregionalism is the core issue in the minds of most Belgians. In fact, one of the frequently noted paradoxes of Belgium is that the administrative structure "accentuates a phenomenon toward which there is much indifference by the masses."[70] Although surveys taken at times of major Fleming-Walloon friction have indicated that the communities problem is paramount in the minds of most Belgians, others conducted in periods of relative ethnoregional tranquility have reported that most Belgians rank community problems well below such issues as the economy and taxes.[71] Moreover, many Belgians do not define themselves primarily in ethnoregional terms. In fact, many of the surveys that have been conducted indicate a relative indifference to community identity on the part of one-third to one-half of the Belgian population,[72] and certain elements of the Belgian system serve to encourage a sense of Belgian identity. Chief among these is the monarchy, and the skill of the present king in heading off several crises between the language communities has earned him wide support throughout Belgium. In addition, important aspects of Belgium's economy are organized nationally, and Brussels, both as capital and as the home of the headquarters of such international institutions as the European Community and NATO, exerts a certain unifying influence.[73] Finally, certain cultural and social institutions continue to bridge the language boundary, and a burst of something close to

---

[70] Georges Goriely, "Frontière linguistique et destin de la Belgique," *Politique Etrangère*, 47, 3 (1982), p. 673. (Author's translation).

[71] Verdoodt, "Les problèmes communautaires belges," p. 3.

[72] See for example, Delruelle and Frognier, "L'opinion publique et les problèmes communautaires," pp. 9; 23; 11.; L'Institut Belge de Science Politique—Projet AGLOP-GLOPO, "Les citoyens belges et leur conception du monde politique," *Res Publica*, 17, 2 (1975), pp. 319-325. Not surprisingly, indifference is around 60 percent in the Brussels area. INUSOP-UNIOP and le Centre de Sociologie Générale, *Bruxelles et sa banlieue: opinions et attitudes des habitants à l'égard des problèmes politiques, linguistiques, sociaux, urbains et culturels, enquête sociologique* (Brussels: Institut de Sociologie, Université Libre de Bruxelles, 1985), p. 34.

[73] A somewhat dated, but insightful, discussion of the factors promoting Belgian unity can be found in Val R. Lorwin, "Belgium: Religion, Class, and Language in National Politics," in *Political Opposition in Western Democracies*, ed. Robert A. Dahl (New Haven: Yale University Press, 1966), pp. 176-178.

widespread Belgian patriotism can accompany the success of the national soccer team in international competition.

It does not follow from these observations, however, that ethnoregionalism is unimportant in modern Belgium. The language regions have become a political and social reality that cannot be ignored. As a result, more Belgians have come to feel at least some sense of ethnoregional identity in the past twenty years than ever before. This is certainly the implication of one of the few surveys that compared changing attitudes over a span of several years. That study concluded that between 1956 and 1968 feelings of Belgian identity had significantly diminished while ethnoregional identity, particularly among Flemings, had grown.[74] Even though the accuracy of surveys such as this can be questioned, the very act of asking questions about ethnoregional affiliation is part of the process of forging collective identity that has accompanied the regional changes of the past few decades.

The strength of collective ethnoregional identity differs greatly among the regions. It is best developed in Flanders, where a sense of the territorial integrity of Flanders as a cultural-political unit is relatively widespread. In interviews with a number of Flemings living close to Brussels, for example, the subjects frequently expressed feelings of resentment over the use of French in local commercial establishments, describing this as an unjustified invasion of their cultural turf. This kind of feeling is given concrete expression when members of the Volksunie pull the emergency brake on trains passing into Flanders if the conductor fails to switch into Dutch. Reactions of this sort are arguably the legacy of a long history of linguistic inequality, but the territorial developments of the twentieth century have given that legacy concrete spatial expression. Among Walloons, by contrast, there is a much less well developed sense of cultural or territorial affiliation.[75] Attempts to forge one are at the heart of the Walloon movement now.[76] Although brought together to some degree by a reaction against Flemish nationalism, shared economic concerns, and linguistic/cultural similarities, the regional restructuring of Belgium and its impact on interaction is arguably the most important force in structuring

---

[74] See, for example, Amédée Kabugubugu and Joseph R. Nuttin, "Changement d'attitude envers la Belgique chez les étudiants flamands," *Psychologica Belgica*, 11 (1970-1971), pp. 27-30, a study comparing attitudes in 1968 with those in 1956.

[75] Quévit attributes this to the internationalist perspective of the dominant Socialist party, the dependence of Walloon intellectuals on France, and the continued strength of local attachments. Michel Quévit, *La Wallonie: l'indispensable autonomie* (Brussels: Entente, 1982).

[76] *Pour une culture de Wallonie* (Petit-Engheim: Yellow Now, 1985); Jean-E. Humblet, "Adresse aux wallons," *Le Soir*, March 27, 1985, p. 1.

and sustaining Walloon collective identity. Brussels presents an extremely stark case of regional change influencing group identity, and whatever collective sentiment may exist in the capital is surely a direct outgrowth of twentieth century political-territorial developments.

Even though collective identity has different roots in the three major language regions, the important point is that legislated regional change is now a primary force in promoting and sustaining group identity in all three. For many of the leaders of the modern Flemish and Walloon movements, the concepts of administrative/political autonomy and group identity have become inextricably interwoven.[77] This is not to suggest that there is a strong separatist sentiment on any side. The pragmatics of Belgian unity have kept separatist ideology on the fringe of Belgian politics.[78] What is important is that Belgians are functioning in a society that is structured to emphasize intra rather than inter-group contact and in an environment in which regional distinctions are a prominent focus of discourse. Under these circumstances, identification at the ethnoregional level is encouraged.

The Question of Mutual Understanding

In a study conducted almost three decades ago, Nuttin found that contact between Flemings and Walloons tended to generate greater tolerance and sympathy.[79] In an environment structured to discourage this kind of interaction and to encourage the formation of ethnoregional identity, one must at least raise the question of its impact on mutual understanding. A 1975 survey of Flemings, Walloons, and residents of Brussels sought to establish the degree of expressed sympathy for inhabitants of each of the three language regions by asking members of each group to rate their level of sympathy on a scale from 0 to 100.[80] Not surprisingly, each group gave a much higher score to those living in the same language region. Sympathy scores for the inhabitants of the other regions were

---

[77] See for example, "Les signataires du manifeste wallon persistent et signent: pas d'autonomie culturelle sans autonomie politique," *Le Soir*, March 8, 1985; "<<Nous voulons plus d'autonomie>>: le leitmotiv de la fête de la communauté flamande," *Le Soir*, July 12, 1985.

[78] As a prominent Flemish politician and member of the Volksunie stated in an interview on July 15, 1986, "I am not Belgian from the heart, but I am a Belgian for practical reasons." (Author's translation.)

[79] Nuttin, "De Ontwikkeling van de Gezindheid Tegenover de Walen en het Persoonlijk Contact," *Tildschrift voor Opvoedkunde*, 5 (1969), pp. 315-333.

[80] AGLOP, *Le citoyen belge dans le système politique*, a data set in the Belgian Archives for the Social Sciences, Louvain-la-Neuve, 1975, reported in McRae, *Conflict and Compromise in Multilingual Societies: Belgium*, p. 109.

markedly lower. Thus, among Flemings the average sympathy score for the people of Flanders was 76, whereas the Flemings had an average sympathy level of 42 for those living in Wallonia and 48 for the inhabitants of Brussels. Among Walloons the figures were 77 for their own region, 49 for Flanders, and 63 for Brussels.[81] To put these figures in perspective, McRae notes that a similar study conducted in Switzerland revealed much weaker interregional differences. It is, of course, impossible to determine how these results would compare with feelings of mutual understanding among previous generations of Belgians, but they do suggest the significance of the present problem, particularly between Flanders and Wallonia.

Certainly with the introduction of strict regional unilingualism outside of Brussels, less tolerance is now shown for those unable or unwilling to use the regional language. A study in the late 1970s revealed that in the Brussels suburb of Sint-Genesius-Rode, the older generation showed a more yielding attitude toward the use of French than the younger generation.[82] It is not unreasonable to suspect that the attitude of the younger generation is suggestive of a stronger rejection of Francophones in general. Moreover, with the organization of the language regions into bases of political power, conflicts over issues ranging from comparative subsidies to the number and quality of cultural events serve to foster interregional antagonisms. Even more importantly, the impact of the new arrangements in reducing the significance of crosscutting cleavages and prompting the division of many institutions along ethnoregional lines serves to reduce arrangements and forums that at one time were sources of mutual cooperation and, by extension, understanding.

The nature of intergroup perceptions is, of course, difficult to uncover. Van Haegendoren captures many of the traditional stereotypes in his account of Flemish and Walloon national characteristics. With the usual disclaimers, he writes of Flemings as calm, deliberate, reliable, apathetic, slow, conservative, traditionalist, inclined to group formation, and attracted to system and method; and the Walloons as lively, hotheaded, impulsive, progressive, individualistic, and inclined toward improvisation.[83] Although these generalizations suffer from all the problems that abstractions of this sort tend to embody, they are representative of widely

---

[81] The significantly greater level of expressed sympathy for the inhabitants of Brussels by Walloons than by Flemings suggests the role of language group affiliation in affecting responses.

[82] Piet Van de Craen and Ann Langenakens, "Verbale Strategieen bij Nederlandstaligen in Sint-Genesius-Rode," *Taal en Sociale Integratie*, 2 (1979), pp. 97-139.

[83] Maurits Van Haegendoren, *The Flemish Movement in Belgium*, translated from the Dutch (Antwerp: Flemish Cultural Council, 1965), p. 10.

shared group views as suggested both by similar accounts from other authors and by surveys of Fleming and Walloon perceptions of one another.[84] With Flemings and Walloons functioning in more autonomous environments, the opportunities for individuals to penetrate beyond these stereotypes may be diminishing. The importance of this point is revealed in the media coverage of an imbalance in the number of Flemish and Walloon youths being sent to detention homes. A leading French-language newspaper ascribed the imbalance to differences in mentality between Flemish and Walloon judges, and pointed to the difficulty of overcoming these differences in view of the separate organization of the judiciary in Wallonia and Flanders and the consequent lack of cross-fertilization of ideas.[85]

To suggest that the present social and political structure of Belgium may be an impediment to the development of mutual understanding is not to argue that the regional developments of the past three decades were a mistake. In fact, they were probably both necessary and inevitable. It is merely to point out that such an approach carries with it certain costs, an understanding of which is necessary to cope with the problems of the future. The regional developments of the twentieth century have created increasingly separate worlds for the Flemings and Walloons. Under these circumstances it is easy for false perceptions and misunderstandings to persist.[86] Thus, among Francophones there is still widespread feeling that the Dutch speakers are trying to usurp most positions of power in the country or to require Francophones who wish to participate in any way in matters of national scope to learn the Dutch language, a tongue that many French speakers view as difficult to learn and of limited importance. Feelings are prevalent among Dutch speakers that the Francophones are out to steal the national capital and as much surrounding territory as possible, and that the Flemings still are not equal partners in all aspects of social, political, and economic life. It is the persistence of such attitudes among Francophones and Dutch speakers that have contributed to the stalemate over Brussels and the controversy over the allocation of national resources, and which consequently present a major challenge for the decades ahead.

There are some indications of growing concern over the negative implications of the now entrenched regional structure of Belgium for

---

[84] See, for example, Fernand Desonay, "Two Ways of Looking at Flanders: A Walloon View," *Delta*, 6, 4 (Winter 1963-1964), pp. 33-39; Nuttin, "Het Stereotiep van Wallen."

[85] "Sur cent jeunes belges placés en homes, nonante-huit francophones! Une fatalité," *Le Soir*, October 19-20, 1985.

[86] A similar point is made by McRae, *Conflict and Compromise in Multilingual Societies: Belgium*, p. 320.

mutual understanding. One is the effort referred to above by the Dutch-language television and radio station to produce programs on Wallonia. Another is a developing exchange program for secondary school students between Flanders and Wallonia.[87] Although the primary objective of the program is to encourage greater proficiency in the other major national language, the potential benefits for mutual understanding are obvious. The need for such initiatives suggests the reality of mutual misunderstanding, but their existence also testifies to a growing recognition of the problem.

The Focus on Territorial Issues

There can be little question that the development of conceptual, functional, and formally instituted linguistic regionalism has served fundamentally to recast community issues and problems in Belgium in territorial terms. In post-World War II Belgium, attention has come to be focused on such matters as the delineation of the language border, the situation of institutions and people located in one region that do not employ the regional language, the status of areas along the language border that do not fall neatly into one region, the appropriate border for Brussels and the status of its surrounding suburbs, the need to divide Brabant so that all provincial boundaries will correspond to the language border, and the proper location for the regional capital of Wallonia. Each of these has become an issue as a consequence of twentieth-century territorial developments.

The central role of territorial considerations is well illustrated by the most recent crisis in the Fourons. In the fall of 1986 the Belgian prime minister offered his resignation over his failure to resolve a dispute arising out of a court ordered dismissal of the Francophone mayor of the Fourons, José Happart, for insufficient knowledge of Dutch,[88] and the following year the government fell over the issue. The incident came about because Happart had been democratically elected in a commune that, as a result of the 1963 language laws, is now officially part of Flanders. The very existence of the issue, then, is a consequence of the regional restructuring of Belgium.

Happart was elected by the Retour à Liège party, an organization that came into being to oppose the political-territorial transfer of the commune to Flanders. He refuses to use Dutch in official meetings and communications, continually asserting his democratic right to follow the will of the voters with respect to language use. In so doing, he is avowedly

---

[87] "Des échanges nord-sud pour les élèves du secondaire," *Le Soir*, October 8, 1984.

[88] For an account of this event, see "Belgium's Leader Offers Resignation," *The New York Times*, October 15, 1986.

seeking to juxtapose the issue of individual rights against that of territorial affiliation. His defense against criticism that his actions are harming the country as a whole is that Belgium is an artificial state, with no political-territorial raison d'être. Even though the Fourons comprises a tiny area with less than 0.05 percent of Belgium's total population, the status of the commune has been the focus of some of the most intense interregional conflicts of the past twenty years, and the declaration of Happart's incompetence is only the latest development undermining the stability of the government. Although the king refused the prime minister's resignation and assigned the matter to a special commission for further investigation, the fact that disputes over this area can generate such far-reaching consequences illustrates dramatically the power and salience of territorial issues in the modern development of the communities problem.

Many other examples could be cited to make the same point. By casting problems in regional/territorial terms, the focus has shifted from social and cultural issues to what Curtis describes as "areas of contact between Flemings and Francophones—Brussels, its periphery, and the language frontier."[89] Problems that would otherwise have no particular ethnoregional significance thus take on a regional dimension. We have already seen this in the case of the school wars and the abdication of King Leopold III after World War II. More recently, such issues as harbor politics and the price of milk have acquired regional/territorial significance.[90] The new game is structured as a territorial one, and the central issue for the future has become the appropriate form of political-territorial organization for the Belgian state.[91]

The Importance of Interregional Comparisons

An extension of the shift in focus to territorial considerations is the heightened salience of interregional contrasts. The extent and visibility of information on regional differences has done much to promote a sense of regional competition in Belgium. The newspapers are filled with comparisons of the number and status of Flemings and Francophones in matters ranging from the foreign service to the Brussels police force. The titles of a few articles of this sort are suggestive: "Linguistic Disequilibrium

---

[89] Arthur E. Curtis, "New Perspectives on the History of the Language Problem in Belgium," (Ph.D. dissertation, University of Oregon, 1971), p. 449.

[90] Léo Tindemans, "Discussion—rapport introductif sur 'Bruxelles et le fédéralisme'," *Res Publica*, 13, 3-4 (1971), pp. 432-433.

[91] See *Waarheen met België?/Que pourrait devenir la Belgique?* (Brussels: Center for the Study of Political Institutions, 1984).

in Airline Employment: Wallonia Protests," "The Walloons have 44,000 Civil Servants Too Many," "Too Many Flemish Civil Servants in Agriculture."[92] Parity issues have also been raised over the composition of a host of institutions, including government bodies, parastatal organizations, the armed forces, the universities, postgraduate research institutions, cultural organizations, public corporations, and the broadcasting networks.

The parity debate is the most intense over issues of government spending. With more than half of the Belgian population, Flanders expects a larger share of government subsidies. Consequently there is considerable resentment over the substantial subsidies that have gone to the ailing Walloon steel industry. Some Walloons counter with the argument that the economic expansion of Flanders was achieved as a result of national cooperation, and that Wallonia deserves nothing less. Given the nature and depth of regional structures, however, national unity on economic issues is increasingly problematic. The usual pattern now is that a decision made to benefit one region is the basis for a campaign to obtain a comparable benefit for the other region. For example, the decision to build an oil refinery in Wallonia was used by Flemish politicians as an argument for the national government to drop opposition to an oil pipeline between Rotterdam and Antwerp.[93] Demands of interregional equilibrium have been a part of decisions ranging from mine closures and rail abandonments to the construction of roads.[94] Interregional competition is particularly ugly in cases in which one region threatens or acts to undercut the other region. Examples include a decision by Flemish industrialists to set up an aircraft construction subcontracting business to compete with a similar successful venture in Wallonia or a threat by Walloon industrialists to use the French port of Dunkirk for exports instead of Zeebrugge unless the government provided Wallonia with greater economic subsidies.[95]

The question of subsidies is important in matters of education, scientific research, and culture as well. Assertions of interregional inequalities have been made in each of these cases, and the budgets for all three have been regionalized. Substantial controversy has surrounded the

---

[92] "Déséquilibre linguistique aux voies aériennes: la Wallonie proteste," *La Libre Belgique*, March 29-31, 1986; "Walen Hebben 44.000 Ambtenaren te Veel," *De Standaard*, November 15, 1984; "Trop de fonctionnaires flamands à l'agriculture: un nouveau dossier linguistique ouvert," *Le Soir*, September 17, 1984.

[93] Hugh Clout, *Regional Variations in the European Community* (Cambridge: Cambridge University Press, 1986), p. 87.

[94] Lode Claes, "The Process of Federalization in Belgium." *Delta*, 6, 4 (1963-1964), pp. 43-52.

[95] See Clout, *Regional Variations in the European Community*, p. 88.

question of relative regional shares given disparities in regional population size and questions of the need for compensation for past imbalances. Compromises have generally involved granting Flanders more than 50 percent of the budget, but controversy persists over a number of specific issues.

The basic point is that the reconstitution of Belgium's political geography along ethnoregional lines has moved issues of relative regional status to the center stage. This helps to explain, for example, the attention paid to comparative regional demographic trends. Walloons have long been concerned about the lower rate of population increase among the Francophones of Belgium. With resource allocation now being tied in many cases to regional population size, the issue has taken on renewed significance, and the focus has shifted from linguistic group trends to regional differences. As a result, the press has given extensive coverage to indications that the gap between the growth rates in Flanders and Wallonia may be narrowing.[96] With relative regional status at the heart of many allocation issues, interregional competition is necessarily a salient issue in contemporary Belgium.

The Emphasis on Regional Problems and Possibilities

The devolution of power to the regions, the division of institutions along ethnoregional lines, and the dichotomy of information have served to focus much attention at the regional level. Considerable power remains with the central government, of course, and Brussels continues to exercise a strong centralizing role in Belgium, but the system is now structured in such a way that some concern with regional affairs is almost unavoidable. This is reflected in the numerous organizations, both public and private, that have come into being to further regional interests. The developing emphasis on the regional level can be seen, on the one hand, as an erosion of the prior dominance of the national level and, on the other hand, as a reflection of an emerging cleavage that is assuming an increasingly important role in Belgian society.

A wide range of matters that once were controlled exclusively at the national level are being transferred to the language communities and regions, and the scope of regional competence and initiative is consequently broadening. A case in point is the recent conclusion of an agreement of economic cooperation between the Walloon region and Quebec.[97]

---

[96] For example, "Reprise en Wallonie...," *Le Soir*, July 9, 1986; "Minder Vlaamse, Meer Waalse Baby's in 1984," *De Standaard*, March 29, 1984.

Moreover, the regional and community councils are attracting considerable media attention by moving decisively into important aspects of domestic policy. The decision of the Flemish cultural council on the use of languages in private business establishments in Flanders is a particularly striking example, as are recent moves by the Walloon regional council to encourage industrial development.[98]

These two examples are suggestive of what is often regarded as a fundamental difference between Wallonia and Flanders in terms of how the region and its status are conceptualized. Verdoodt characterizes the difference as an emphasis among Walloons on decentralization and economic development, in contrast to a basic concern among Flemings with the maintenance and development of their culture.[99] There is support for such a generalization beyond the legislative initiatives mentioned above. On the Walloon side, the regional council issues public relations literature for foreign distribution touting Wallonia's tourist attractions and its advantages as a place for investment, and containing broad statements about Wallonia as a region that is master of its own future.[100] Much of the literature on Flanders and the Flemish cause continues to focus on issues of language, culture, and sense of community.[101] These differences in orientation may in part be a legacy of the bases of Flemish and Walloon regionalism during the twentieth century. The important point is that the emphases are different. As a consequence, solutions to problems are increasingly being sought at the regional rather than the national level.

The significance of the gradual shift during the twentieth century to a focus on the language region is that it has served to alter, and in some cases largely to supplant, historical cleavages in Belgium. All of the major social divisions in Belgium—language, politics, religion, and class—have been

---

[97] "La région wallonne et le Québec, partenaires privilégés," *La Libre Belgique*, April 21, 1986.

[98] "Wathelet et Morland tentent de créer la Walloon Valley," *Le Soir*, July 3, 1986.

[99] Verdoodt, *Les problèmes des groupes linguistiques en Belgique*, pp. 55-56.

[100] *Wallonie* (Liège: Massoz, n.d.). The French-language community issues a similar periodical covering both Wallonia and Brussels. *Wallonie—Bruxelles: une communauté/deux régions*, revue trimestrielle éditée par le Commissariat Général aux Relations Internationales de la Communauté Française de Belgique (1983-1986).

[101] The vast majority of entries in the section entitled "Cultural Problems" of the most recent bibliography on language and society in Belgium are about Flanders and the Flemish. Lieve Suenaert and Pierre Verdoodt, *Langue et société en Belgique 1980-1985: bibliographie analytique et guide du chercheur/Taal en Maatschappij in België 1980-1985: Analytische Bibliografie en Gids voor de Gebruiker*, unpublished volume of the Centrum voor de Studie van de Pluriculturele Maatschappij/Centre pour l'Etude de la Société Pluriculturelle, 1986, section 16.

affected to some degree by the growth and institutionalization of regionalism. Linguistic issues are now only the backdrop for territorial disputes. With the rise of strict regional unilingualism outside of Brussels, language community and regional interests increasingly overlap. In fact, this has led to the merging of the Flemish community and region, and there is a growing movement among Francophones to follow suit. In reality, some of the French-speaking community and regional institutions are already working together, although the future of merger proposals are as yet uncertain.

The impact of regional developments on crosscutting political cleavages has already been mentioned. With the splitting of the traditional parties and the division of associated political institutions along ethnoregional lines, patterns of elite cooperation within ideological families are being substantially eroded. Representatives of the major political parties frequently do not vote together on issues of dispute between the language communities.[102] The tendency for patterns of support for particular issues to break down along regional/community lines rather than party lines is further encouraged by a structure in which regional executives and representatives are also involved at the national level. [103] In a system in which political/ideological structures have long been the major force in government, this is a significant development.

Religion and class differences still apply to the whole of Belgium, but even in these areas regionalism has had an impact. There is no strict correlation between religious practice and language region, but the tendency to look at aggregate differences between the regions has led to the strong association of clericalism with Flanders and anti-clericalism with Wallonia.[104] As for class conflict, despite many examples of working-class solidarity from the history of Belgium, regional divisions overshadowed class consciousness in the dispute over the general strike of 1960-1961, and the resulting divided internal structure of the labor unions now helps to sustain regional divisions. Moreover, the regionalized economic structures, the formal impediments to extraregional investment, and the different

---

[102] James A. Dunn, *Social Cleavage, Party Systems and Political Integration: A Comparison of the Belgian and Swiss Experiences*, (Ph.D. dissertation, University of Pennsylvania, 1970), p. 272.

[103] See, for example, "Les blocs flamands et francophones s'affrontent à la chambre autour de la périphérie," *Le Soir*, May 5, 1984; Spitaels - Van Miert: du duel communautaire au duo socio-economique," *Le Soir*, October 22, 1984.

[104] The data support such a generalization, of course, but, in keeping with the nineteenth century patterns documented in chapter 4, if levels of religiosity are evaluated independently of the linguistic distributions, a pattern emerges that does not reflect closely the language regions.

approaches being taken to regional development in Flanders and Wallonia discourage the development of a unified industrial class.

In short, the institutionalization of regional difference is far more than a simple reflection of territorially based intergroup tensions. It profoundly alters the ways in which problems are conceived, relationships are understood, and solutions are sought. In forging a strongly regionalized state, Belgium's political elite has solved certain problems, but has created others as well. In the process, the nature and context of ethnoregionalism has been fundamentally remade.

Chapter 8

CONCLUSION

The changing nature of regional arrangements in Belgium lies at the heart of the development and current character of the communities problem. By focusing on the dynamics and implications of regional formation and change, this study has sought to shed some light on the significance of conceptual, functional, and formal compartmentalizations of space for the evolution of ethnoregionalism. Implicit in this effort has been an attempt to understand regionalism as more than a simple characterization of a variety of substate nationalism in which a group seeks to assert control over a particular area within a state. Instead, regionalism is seen as a process that reflects and structures patterns of group identity and interaction. In Belgium, looking at why and how regional divisions have come to correspond to ethnolinguistic patterns, and the consequences of that process, yields insights into the role of territory and space in the substate nationalist context beyond straightforward distributional considerations.

Four interrelated ideas have provided the backdrop for this study. They are: (1) that regional divisions along ethnic or cultural lines are not natural or inevitable, but arise out of particular historical and geographical contexts; (2) that present regional divisions cannot necessarily be projected into the past in our efforts to explain the rise of ethnoregionalism; (3) that ethnoregionalism is not a mechanistic response to territorial variations in the distribution of particular phenomena; and (4) that regional divisions are not only reflections of spatially significant ethnic or cultural tensions, but are important forces in shaping group interaction patterns and intergroup issues and concerns. Rather than taking present regional arrangements for granted, these themes draw attention to important aspects of the territorial dimension of cultural pluralism in Belgium.

CONCLUSION

The communities problem did not always have a strong territorial foundation. There was no well developed sense of regional ethnolinguistic identity in Belgium at the time of independence in 1830, and the present regions of Flanders and Wallonia had no functional or political integrity. Throughout the nineteenth century, debate focused primarily on individual language rights and the status of linguistically defined groups rather than on specifically regional concerns. Moreover, the fact that Flanders and Wallonia did not constitute meaningful economic or social territorial units before the twentieth century, casts doubt on explanations of the rise of the Flemish movement that are premised on generalized assumptions of regional inequalities. This is not to suggest that Flemish nationalism did not have a social or economic basis. It most certainly did. However, the Flemish movement did not develop as a regional response to social and economic conditions in northern Belgium, but as a reaction among intellectuals in a few major northern cities to the disadvantaged position in the Belgian state of Flemings who did not speak French. In fact, a primary concern of the leaders of the movement was to spread this message to the wider Flemish population.

The shift to a more specifically territorial orientation among the leaders of the Flemish movement during the first third of the twentieth century grew out of complex circumstances. The most important underlying factors were the failure of the nineteenth-century laws guaranteeing limited individual linguistic rights to bring about significant change; the desire to find an alternative to a system that, because of suffrage requirements, put the Flemings as a group at a disadvantage; the need for a unifying focus for group identity in the North; and the impact of an imposed administrative division of Belgium by the Germans during the occupation of World War I. Many of the same concerns led the leaders of the small, reactive Walloon movement to press as well for a degree of regionalization. The period since World War I has seen the growth, acceptance, and institutionalization of many of these ideas, culminating in the restructuring of the Belgian state and the devolution of power to the language regions. In the process, functional patterns have come to correspond more closely to ethnoregional divisions, patterns of group interaction have been fundamentally altered, and territorial issues have moved to the center of the communities debate.

Linguistic differences have been the backdrop for the communities debate throughout Belgian history. To describe the situation as a language problem per se, however, is misleading.[1] The Flemish movement initially

---

[1] O'Barr points out that the real issues are rarely linguistic, in that language usually becomes an issue as a consequence of differentials in power relations. William M. O'Barr, "The Study

developed in reaction to linguistic inequalities, and language has remained the central symbolic focus of the movement. The embracing of wider social and economic issues affecting the Flemings as a whole gave the movement a strong social character by the turn of the twentieth century. During the course of this century both the Flemish and Walloon movements have come to be centrally concerned with regional issues. Although language is one of these issues, it is part of a wider debate over relative regional status and interregional interaction. To a significant extent, language group membership has been replaced by regional affiliation as the primary focus of group identity in Belgium.

Although a large number of states are confronted with internal ethnoregional divisions, the Belgian case is, in many ways, unique. The relative numerical and territorial parity between groups is found in few other states. In addition, aside from language, the cultural differences between Flemings and Walloons are relatively minor, a product of long historical associations that bridged language boundaries. There is a widespread commitment to the resolution of problems within an essentially democratic framework, which is reflected in one of the most complex sets of laws related to internal ethnoregional differences of any country in the world. Moreover, despite interregional differences in prosperity within Belgium, the vast majority of Belgians enjoy a standard of living that is among the highest in the world. Finally, the recency of regional developments themselves serve fundamentally to distinguish Belgium from many other cases.

The importance of the last point is revealed by the fundamental differences between the historical character of territorial divisions in Belgium and Switzerland. Comparisons between these two cases are problematic because Switzerland, unlike Belgium, developed with firmly rooted territorial expressions of ethnic differences in place. In fact, "the *principle of territoriality* has been enshrined in Swiss tradition since long before the creation of the modern Confederation in 1848."[2] This has meant long-standing acceptance of behavioral norms that include the expectation that everyone will deal with local authorities in the regional language, that all communications to those authorities from the federal government will be in that language, and that the regional tongue will be used in the schools. Although the legislative and constitutional changes in Belgium over the past three decades have introduced similar formal arrangements in that

---

of Language and Politics," in *Language and Politics*, ed. William M. O'Barr and Jean F. O'Barr (The Hague: Mouton, 1976), pp. 1-27.

[2] James A. Dunn, "Social Cleavage, Party Systems and Political Integration: A Comparison of the Belgian and Swiss Experiences," (Ph.D. dissertation, University of Pennsylvania, 1970), p. 30.

country, to assume that Belgium and Switzerland are now fundamentally alike is to ignore the historical context.

Of profound importance to an understanding of the Belgian case is the recognition that throughout much of the twentieth century the country has been confronted with an intense struggle over regional linguistic issues, with all that this implies in terms of social upheaval and intergroup tensions. In the process, Belgium has gone through a wrenching change from a highly centralized state to a regionalized political entity, and the devolution of power to the regional level is probably not complete. The significance of not having a strong tradition of distinct and widely accepted language regions is suggested in a tabulation made by Dunn of the number of articles on language issues and reports of language-related protests in the Francophone presses of Belgium and Switzerland over separate three month periods in 1969. He found 14 reported incidents of language protest and 107 articles on language issues in the Belgian press as opposed to no reports of language protest and only 15 articles related to language matters in the Swiss papers.[3] Although many Belgians hope that the legislative and constitutional changes of the past thirty years will lay the groundwork for the development of a stable and widely accepted regional ethnolinguistic structure for the country, even under the best of circumstances this will take much additional time.

Given the distinctiveness of the Belgian case, the significance of a focus on regional/territorial developments does not lie in the potential transferability of the Belgian experience to other contexts. Rather, it is found in the insights that can be gained from an incorporation of the regional dynamics of cultural difference into the analysis of substate nationalism. Many states have adopted, or are contemplating the adoption of, special territorial arrangements in response to internal ethnoregional conflicts. With states as diverse as Nigeria, Sri Lanka, and Spain in this category, no one country's experience with ethnoregionalism is likely to be particularly suggestive for another. What may be of more general significance, however, is an appreciation and evaluation of issues that proved to be revealing in the Belgian case. These include the nature and historical context of regional divisions, the spatial correlation between cultural and other social and economic patterns, the nature and role of group territoriality, and the legacy of social and institutional compartmentalizations of space along cultural or ethnic lines.

The creation of politically and functionally significant regional units can be an effective, even necessary means of coping with an immediate

---

[3] *Ibid.*, p. 262. The papers used for the study were *La Libre Belgique* and *Le Soir* in Belgium. In Switzerland Dunn consulted *La Suisse* and *La Tribune de Genève*.

problem. One of the most important of the lessons that emerge from this study, however, is that territorial arrangements are rarely final solutions to problems of internal discord because they do not address the underlying causes of those problems. To make this point is not to advocate the perpetuation of centralized, unitary government. In fact, it can well be argued that the devolution of power along ethnoregional lines provides the best hope for reducing internal tensions.[4] The suggestion offered here is that formal regional arrangements both confront and create problems.

In general terms, the significance of institutionalized territorial arrangements is that they require at least some form of adjustment for those who do not conform to established regional norms.[5] An extension of this is that they tend to discourage interaction between formal territorial units. Moreover, by recreating the context of group interaction and identity, formally instituted regional arrangements frequently cast the issues of contention in a different form. They can serve to encourage territorial tensions because they impose static spatial arrangements on essentially dynamic processes (for example, the dispute over the expansion of Brussels).[6] The enduring legacy of territorial arrangements can be seen in the German and Swiss cases where, as McRae points out, the imprint of religious territoriality is visible to this day even though both countries now recognize religious freedom on an individual basis.[7] This suggests the importance of an appreciation of the problems and potentials of territorial arrangements as responses to internal ethnic or cultural discord.

Unfortunately, most general treatments of the Belgian case devote only scant attention to the territorial dimension of the communities problem. Reference is often made to the adoption of a territorial strategy by the Flemish movement in the early twentieth century and the regional character of the recent language laws, but little has been written about the reasons for regional developments or their implications. This study has sought to move into this void by identifying and analyzing factors

---

[4] Thus, Cavin concluded that nonterritorial approaches have not generally been successful in coping with substate nationalism. Jean-Françoise Cavin, *Territorialité, nationalité et droits politiques* (Lausanne: Held, 1971), chapters IV-VI.

[5] See generally, H. Kloss, "Territorialprinzip, Bekenntnisprinzip, Verfügungsprinzip: Über die Möglichkeiten der Abgrenzung der Volklichen Zugehörigkeit," *Europa Ethnica*, 22 (1965), pp. 52-73.

[6] This point is made in connection with the Arab-Israeli conflict by Manoucher Parvin and Maurie Sommer, "Dar Al-Islam: The Evolution of Muslim Territoriality and Its Implications for Conflict Resolution in the Middle East," *International Journal of Middle East Studies*, 11, 1 (February 1980), pp. 1-21.

[7] Kenneth D. McRae, "The Principle of Territoriality and the Principle of Personality in Multilingual States," *Linguistics*, 158 (August 1975), p. 36.

precipitating the conceptual, functional, and formal regionalization of Belgium along ethnolinguistic lines and the ways in which regional change has shaped the evolution of the communities problem.

The concepts of territoriality and regionalism provide essential insight into these issues. A view of regions as social creations that become an integral part of the context for social change allows us to move beyond static conceptualizations of substate nationalist conflict based primarily on distributional or functional considerations, and to address how and why particular spatial patterns acquire significance in human affairs. The concept of territoriality provides insight into the ways in which regions are created in ethnically divided societies, focusing attention on an important means by which group leaders seek to achieve nationalist objectives and governments attempt to cope with internal conflict. Indeed, the activities of the Flemish and Walloon movements, and of the Belgian government, provide concrete evidence of many of Sack's tendencies of territoriality outlined in chapter 2.[8] Regionalism and territoriality, then, are important additions to the geographer's conceptual tools for the investigation and analysis of substate nationalist issues.

Once the Belgian communities debate was cast fundamentally in terms of regional problems and concerns, the adoption of territorial language laws was almost unavoidable. It was the only realistic way for Belgium's political elite to resolve the conflict without losing their position or their country.[9] On one level the legislated regional developments of the twentieth century are remarkable in that they have served to maintain Belgian unity without significant violence. On the other hand, institutionalized ethnoregionalism has created new problems, many of which were probably not foreseen at the time of implementation.

The benefits and disadvantages of legislated regionalism in Belgium can be seen in the successes and shortcomings of the regional reforms of the past three decades. These reforms have gone some way toward the satisfaction of the demands advanced by the Flemish and Walloon movements during the first half of the twentieth century. Regional unilingualism has been essentially guaranteed outside of Brussels, provisions are in place to ensure linguistic equality in the capital, and the

---

[8] Robert D. Sack, *Human Territoriality: Its Theory and History*, Cambridge Studies in Historical Geography (Cambridge, England: Cambridge University Press, 1986), pp. 31-42.

[9] This point was made by Covell in connection with the enactment of the early 1960s language laws. Maureen Covell, "Ethnic Conflict and Elite Bargaining: The Case of Belgium," *West European Politics*, 4, 3 (October, 1981), pp. 210-211. The unwillingness of the political class to relinquish power is suggested by the slow pace of legislated regional reform and the persistence of an arrangement whereby members of the national Parliament also sit on the community and regional councils.

language communities have been granted a degree of control over their own affairs. Although significant changes have often been preceded by periods of heightened passion, serious chaos has been avoided. This does not mean that all problems have been solved. Indeed, the communities issue has been at least partially transformed into an ethnoregional issue. This is reflected in the present salience and volatility of such issues as the status of Brussels, the distribution of benefits among regions, and the conflict between democratic and territorial principles in the Fourons.

The impact of the new arrangements on the evolving regional structure of Belgium has been substantial. In addition to the specifically regional ramifications outlined in this study, the country as a whole has paid a relatively high price for regionalization, in terms of both monetary costs and perceptions of government efficiency. The duplication of so many political and governmental structures has placed substantial demands on tax revenues.[10] Although impossible to quantify, the burdens of regionalization have almost certainly contributed to a public deficit that, on a per capita basis, is one of the highest in the world. Moreover, the financial costs of complying with regional language laws can be very high for individual businesses.[11]

Feelings of frustration or cynicism toward the government may also be attributed to some degree to the regionalization process. Debate over regionalization has dominated the Belgian political scene for extended periods on a number of different occasions during the past three decades, and the compromises that have eventually been implemented have often been perceived as unsatisfactory by all groups. Moreover, as Velimsky notes in a study of the future of regionalism in Belgium, "the hair-splitting discussions have reached a degree of significance unimaginable in other multinational states."[12] In view of the incomplete nature of the most recent round of reforms in 1980, this is likely to continue.[13] Although perhaps necessary to the process of moving toward workable solutions, the

---

[10] See generally, Tim Dickson, "High Cost of Cultural Divide," *Financial Times*, June 13, 1986.

[11] Curtis notes that some of the regulations "when followed to the letter, cause considerable inconvenience and inefficiency, with few compensating benefits." Arthur E. Curtis, "New Perspectives on the History of the Language Problem in Belgium" (Ph.D. dissertation, University of Oregon, 1971), p. 558.

[12] Vitezslav Velimsky, "Belgium of the Eighties: Unitary, Bi-Cultural or Made Up of Three Regions?" *Europa Ethnica*, 40, 1 (1983), p. 1.

[13] Jan Ceuleers, "De Staatshervorming van 1980 als Niet-Oplossing," *Res Publica*, 26, 3 (1984), pp. 293-301.

seemingly endless debates, committee reports, and conferences have done little to inspire confidence in government efficiency.

The prolonged and highly publicized battles over ethnoregional issues in the government have led many Belgians to feel that the communities problem is largely a creation of politicians. Such a view does not take into consideration the historical and social roots of ethnolinguistic identity, but it does point to an important facet of the Belgian situation: that the regional changes of the past thirty years have been driven by legislative and constitutional enactments crafted and implemented by politicians. These changes have served to create a regional reality that is of importance even for those who profess indifference toward the communities problem and no strong sense of ethnoregional identity.

The singer Jacques Brel, one of Belgium's native sons, once expressed in song his sense of cultural priorities for his fellow countrymen:

> If I were king I would send all the Flemings to Wallonia and all the Walloons to Flanders for six months . . . They would live with a family and would solve all our ethnic and linguistic problems very fast. Because everybody's tooth aches in the same way, everybody loves their mother, everybody loves or hates spinach. And those are the things that really count.[14]

Of course they do. But the territorial approach followed in dealing with Belgium's communities problem has by now assumed an irreversible importance in national life, all feelings of solidarity aside. The fundamental challenge facing the Belgian state in the decades ahead is to sustain a sense of common purpose in a system structured to discourage interregional interaction and to promote identification at the regional level.

---

[14] Quoted in English translation in "It's Hard Going: A Survey of Belgium," *The Economist*, February 22, 1986, p. 17.

CHRONOLOGY

1830	Belgium gains its independence. Provisional government issues decrees making French the language of command in the military and the language of the official law reporter, and abolishing chairs in Dutch language and literature at the state universities.
1831	The Belgian Constitution establishes freedom of language choice.
1835	Issuance of a royal decree stipulating that lectures in institutions of higher learning be in French.
1839	Publication by Hendrik Conscience of *The Lion of Flanders*, a symbol of the early Flemish movement.
1840	A petition is submitted to Parliament calling for the use of Dutch in public affairs and in education, marking the first major political initiative of the nascent Flemish movement.
1845	Belgian government agrees to provide translations of certain laws for the Flemish communes.
1848	The term Wallonia first appears in print.
1856	Formation of an official commission to investigate measures to regulate the use of Dutch in public affairs and to promote Flemish literature.
1858	Founding of the first Walloon literary organization, the Société Liègeoise de Littérature Wallonne.
1859	Publication of the report of the 1856 commission, which became the manifesto of the Flemish movement throughout much of the nineteenth century.

CHRONOLOGY

1863	For the first time, an oath of office is taken before Parliament in Dutch.
1873	Enactment of the first major piece of language legislation, a law calling for trials in the Flemish communes to be held in Dutch unless the defendant requests French.
1876	Enactment of a law allowing degrees to be granted in Flemish literature at the University of Ghent.
1878	Enactment of a law requiring public notices to be translated into Dutch in the Flemish communes.
1883	Enactment of a law stipulating that in the Flemish part of Belgium, courses in the preparatory section of public secondary schools be conducted in Dutch.
1885	First appearance of parts of the Belgian law reporter in Dutch translation.
1886	Enactment of a law requiring aspiring military officers to demonstrate some knowledge of Dutch. The first Walloon political organization is founded in St. Gilles.
1888	For the first time a speech is made in Parliament in Dutch.
1889	Amendment of the 1873 law on language in the courts to provide for the exclusive use of Dutch in criminal proceedings in the Flemish communes.
1890	Enactment of a law requiring candidates for judicial posts in the Flemish provinces after January 1, 1885, to demonstrate a knowledge of Dutch. The first major Walloon congress is held.
1891	Enactment of a law obliging the appellate courts in Brussels and Liège to hear cases in Dutch on appeal from courts in the Flemish communes. Issuance of a list of Flemish communes for purposes of administering the 1889 law on language in the courts.
1892	Founding of the influential Walloon journal, *La Défense Wallonne*.

CHRONOLOGY

1893	Introduction of universal male suffrage with multiple voting for property owners.
1897	Enactment of a law extending the 1889 law on language in the courts to military tribunals.
1898	Enactment of a law requiring publication of all future laws in both French and Dutch.
1901	Discovery of coal in Limburg.
1902	Replacement of the Flemish literary review *Van Nu en Straks* (From Now and the Future) with a new journal.
1906	Lodewijk De Raet publishes his influential treatise *Over Vlaamse Volkskracht* calling for Dutch to be the official language at the University of Ghent. Episcopate declares French as the sole language of instruction at the Catholic University of Louvain.
1909	Issuance of a report on language at the University of Ghent by a government committee headed by De Raet, calling for a gradual transition to the exclusive use of Dutch at the university.
1910	Enactment of a law requiring university applicants from secondary schools in the Flemish part of the country to be examined in Dutch. Presentation of the first Walloon-oriented list of candidates for parliamentary elections.
1912	Jules Destrée sends a highly publicized letter to the king declaring that there are no Belgians, only Flemings and Walloons.
1913	Enactment of a law requiring students in military school after January 1, 1917, to demonstrate at least some knowledge of Dutch.
1914	Enactment of a law requiring attendance at school for all children under the age of twelve. Germans invade and occupy Belgium.
1916	The occupation government issues a number of decrees promoting the use of Dutch in education and in public affairs.

CHRONOLOGY

1916	Formation of an organization of Flemish soldiers embittered over incidents of language discrimination, which later became the Front Party (Frontpartij).
1917	Founding of the Council of Flanders, which cooperated with the Germans in hopes of achieving greater autonomy for Flanders. Initiation of an administrative partitioning of Belgium along language lines by the occupying government.
1918	The period of German occupation ends and the administrative partition is abolished.
1919	Abolition of multiple voting for property holders. Members of the Frontpartij are elected to Parliament, marking the first time members of a party purporting to represent language community interests enter the national government. Dramatic trial and sentencing of the Flemish wartime collaborator, August Borms.
1921	Enactment of a law stipulating that Dutch be used in the conduct of state, provincial, and communal affairs in the Flemish provinces and districts, marking the first important formal recognition of the language regions.
1923	Issuance of a royal decree calling for Dutch to be the language of administration at the University of Ghent.
1928	Establishment of separate regiments for Dutch and French speakers in the Belgian army. August Borms, in prison for collaboration during World War I, is elected to Parliament from Antwerp.
1929	A general amnesty is granted to wartime collaborators, but no property is returned to those who had been convicted. Founding of the federalist Vlaamsch Studiecomitié voor Politieke, Sociale en Cultureele Aangelegenheden.
1930	Completion of the transformation of the University of Ghent into a Dutch-language institution.

CHRONOLOGY

1932	Enactment of laws calling for complete unilingualism in the Flemish and Walloon parts of Belgium and the establishment of a boundary between the language regions that was subject to alteration in response to changes in language patterns as revealed by the results of the decennial census.
1935	Publication in *De Standaard* of a widely publicized Flemish plan to establish the language boundary definitively, to limit the expansion of Brussels, and to split government services along language lines.
	Enactment of a law instituting sanctions for violation of the requirement that the regional language be used in the courts.
1937	Issuance of an influential report by a private group looking at state reform that rejects federalism.
	Flemish Socialists endorse a plan to bring about conformity between administrative and linguistic boundaries.
1938	Enactment of a law reinforcing the principle of dual unilingualism in the armed services.
1940	Germans invade and occupy Belgium for the second time.
	Germans provide complete reparation to Belgian collaborators who had been prosecuted after World War I.
1944	Defeat of the Germans and reinstatement of the Belgian government.
1945	At the National Walloon Congress, the delegates opt for union with France in a *vote de coeur*, and register approval for a federal structure for Belgium in a *vote de raison*.
1948	Founding of the Harmel Center "to investigate the problems of the Flemish and Walloon regions."
1950	In a referendum over the return of King Leopold III, a negative vote by a majority of Walloons blocks his reinstatement despite a national majority to the contrary.
1952	Introduction of a joint Flemish-Walloon proposal for federalism in Parliament.
1954	Founding of the Flemish federalist party, the Volksunie.

CHRONOLOGY

1954	Publication of the 1947 census figures on language, showing the need for adjustments in the language boundary.
1958	Issuance of the report of the Harmel center, laying the groundwork for later reforms. Conclusion of an agreement ending the "school wars," paving the way for ethnolinguistic issues to move to the center stage.
1960	General strikes in Wallonia, extending into 1961, become an important symbol for growing Walloon ethnolinguistic identity. Removal of language questions from the census following significant protests.
1961	Founding of the Walloon federalist party, the Rassemblement Wallon. The Flemish movement organizes a large march on Brussels to dramatize its cause.
1962	A second great Flemish march on Brussels takes place. Enactment of a law establishing definitively the boundary between Flanders and Wallonia.
1963	Enactment of a law dividing the Brussels district into a bilingual capital district, a special district consisting of six communes with language facilities for Francophones, and a unilingual Flemish district. Enactment of laws adapting the educational system and judicial districts to the regional language boundary.
1964	Founding of the Front Démocratique des Francophones, a party representing primarily the interests of French speakers living in and around Brussels.
1967	Recognition of an official Dutch-language version of the Constitution.
1968	The Catholic University of Louvain is divided into French and Dutch-speaking sections following significant protests, and plans are made for the departure of the French section to a location in Wallonia within a few years.
1969	Division of the Free University of Brussels along language lines.

CHRONOLOGY

1970	Constitutional revisions restructuring the Belgian state along regional language lines. Enactment of a law establishing economic councils for Wallonia, Flanders, and Brabant. Enactment of a law eliminating the special district status of the communes around Brussels.
1971	Enactment of a law dividing Parliament into language groups.
1973	Issuance of a decree by the Dutch community council calling for the use of Dutch in communications between all employers and employees in Flanders.
1977	The political parties in power agree to a plan calling for the devolution of power to the community and regional councils, providing the basis for the 1980 constitutional revisions.
1979	Enactment of a law extending the competence of the community councils to such matters as public health and scientific research, and establishing the power of the regional representative bodies over such matters as employment, housing, water, and energy.
1980	Adoption of constitutional revisions and laws specifying and expanding the powers of the community and regional political institutions.
1982	Issuance of a decree by the French community council calling for the use of French in business establishments in the French-speaking region or in communications with French-speaking workers in any part of Belgium.
1986	Court-ordered modification of the decrees of the French and Dutch community councils regarding language practices in private business establishments. The government offers to resign over its inability to resolve problems associated with the election of a French speaker as mayor of the officially Flemish Fourons commune.
1987	Fall of the Belgian government over the Fourons issue.

BIBLIOGRAPHY

Primary Sources

Official Documents and Publications

Annales Parlementaires de Belgique—Chambre des Représentants, volumes from various sessions. Brussels, November 1844-1986.

Annales Parlementaires de Belgique—Sénat, volumes from various sessions. Brussels, November 1844-1986.

Annuaire de Statistiques Régionales. Brussels: Institut National de Statistique, 1976-1986.

Annuaire Statistique de la Belgique, various volumes. Brussels: Ministère de l'Intérieur or Ministère des Affaires Economiques, Institut National de Statistique, 1870-1986.

Bulletin des Arrêtés du Gouvernement Provisoire. Brussels, October 1— December 7, 1830. Later released as *Moniteur Belge.*

Bulletin Officiel des Décrets du Congrès National de la Belgique et du Pouvoir Exécutif. Brussels, December 8, 1830—September 15, 1831. Later released as *Moniteur Belge.*

Bulletin Officiel des Lois et Arrêtés de la Belgique. Brussels, September 16, 1831—February 24, 1845. Later released as *Moniteur Belge.*

Centre de Recherche pour la Solution Nationale des Problèmes Sociaux, Politiques et Juridiques des Diverses Régions du Pays. *Rapport Final.* Issued as *Documents Parlementaires—Chambre des Représentants— 1957-1958,* April 24, 1958, doc. 940.

Commission Flamande: Installation, Délibérations, Rapport, Documents Officiels. Brussels: Korn, Verbruggen, 1859.

Documents Parlementaires—Chambre des Représentants, volumes from various sessions. Brussels, 1831-1986.

Documents Parlementaires—Sénat, volumes from various sessions. Brussels, 1831-1986.

Gesetz- und Verordnungsblatt für die Okkupierten Gebiete Belgens, various issues. Brussels, February 1916—December 1917.

Gesetz- und Verordnungsblatt für Flandern, various issues. Brussels, January 1918—June 1918.

Gesetz- und Verordnungsblatt für Wallonien, various issues. Brussels, January 1918—June 1918.

Moniteur Belge. Brussels, February 25, 1845—December 31, 1938.

Moniteur Belge/Belgisch Staatsblad. Brussels, 1939-1987.

Pasinomie: Collection Complète des Lois, Arrêtés et Règlements Généraux qui Peuvent Etre Invoqués en Belgique. Brussels: Emile Bruylant, 1833-1986.

Rapport Annuel de la Commission Permanente de Contrôle Linguistique, various issues. Brussels: Commission Permanente de Contrôle Linguistique, 1966-1985.

Recensement de la Population: 31 Décembre 1970. 13 vols. Brussels: Institut National de Statistique, 1975.

Recensement Général (31 Décembre 1856): Population. Brussels: Ministère de l'Intérieur, 1861.

Recensement Général de la Population au 31 Décembre 1947. 11 vols. Brussels: Ministère des Affaires Economiques, Institut National de Statistique, 1949-1954.

Recensement Général du 31 Décembre 1920: Population. 3 vols. Brussels: Ministère de l'Intérieur et de l'Hygiène, 1925-1926.

Résumé de la Statistique Générale de la Belgique. Brussels: Département de l'Intérieur, 1852.

Revue du Travail. Brussels: Ministère de l'Industrie et du Travail, 1896.

Revue du Travail 1900. Brussels: Ministère de l'Industrie et du Travail, 1900.

Statistiques du Mouvement de la Population de l'Etat Civil en 1890. Brussels: Ministère de l'Intérieur et de l'Instruction Publique, 1895.

Verordnungsblatt des Militärbefehlshabers in Belgien und Nordfrankreich für die Besetzten Gebiete Belgiens und Nordfrankreichs, various issues. Brussels, 1940-1944.

General Published Statements, Documents, and Reports

Les archives du Conseil de Flandre (Raad van Vlaanderen). Brussels: Ligue Nationale pour l'Unité Belge, 1928.

Congrès de Concentration Wallonne—Liège—27-28 septembre 1930: compte-rendu officiel. Liège: Wallon Toujours, 1930.

"Le congrès de Liège des 1 et 2 octobre 1949." *Nouvelle Revue Wallonne*, 2, 2 (January 1950), pp. 1-13.

Le congrès de Liège des 20 et 21 octobre 1945: débats et résolutions. Liège: Editions du Congrès National Wallon, 1945.

Congrès Wallon: compte-rendu officiel. Liège: Mathieu Thône, 1905.

Destrée, Jules. "Lettre au Roi sur la séparation de la Wallonie et de la Flandre." *Revue de Belgique* (August 15—September 1, 1912), pp. 735-758.

Economie Wallonne. Rapport présenté au gouvernement Belge par le Conseil Economique Wallon le 20 mai 1947. Brussels: Conseil Economique Wallon, 1947.

Humblet, Jean-E. "Adresse aux Wallons." *Le Soir*, March 27, 1985, p. 1.

Huyttens, Emile. *Discussions du Congrès National de Belgique, 1830-1831*. Brussels: Société Typographique Belge, 1844-1845.

Jennissen, Emile, ed. *Wallons—Flamands: pour la séparation politique et administrative*. Liège: Charles Desoer, 1911.

Ligue d'Action Wallonne. Collection of pamphlets in the Bibliothèque Royale Albert 1er, Brussels, 1926-1938.

Oukhow, Catherine. *Documents relatifs à l'histoire de la première internationale en Wallonie.* Cahier 47. Leuven: Centre de l'Histoire de la Première Internationale en Wallonie, 1867.

Pour une culture de Wallonie. Petit-Engheim: Yellow Now, 1985.

Pro Flandria Servanda. *Flanders' Right and Claim for Autonomy Formulated, Explained, Justified.* The Hague: Martinus Nijhof, 1920.

La réforme de l'état. Brussels: Centre d'Etudes pour la Réforme de l'Etat, 1937.

Rudiger [pseudonym for A. Wullus]. *Flamenpolitik.* Brussels: Rossel, 1922.

Société Liègeoise de Littérature Wallonne. *Cinquantième anniversaire de l'indépendance nationale.* Liège: H. Vaillant-Carmanne, 1880.

"Statut et règlement." *Bulletin de la Société Liègeoise de Littérature Wallonne,* 1 (1858), pp. 5-10.

Union pour la Défense de la Langue Française à l'Université de Gand. Ghent: A. Vanderwerghe, 1910.

Vlaamsche Hoogeschool Commissie. *Verslag over de Vervlaamsching der Hoogeschool van Gent.* Ghent: Algemeen Nederlandsch Verbond, 1909.

Miscellaneous Printed Collections

L'action Socialiste: lettres d'Emile Vandervelde 'à la Gazette de Bruxelles.' Ghent: Volksdrukkerij, 1911.

Houtman, W. *Vlaamse en Waalse Documenten over Federalisme.* Schepdaal, Holland: Het Pennoen, 1963.

Maes, L. Th., and Santbergen, Van, eds. *Documents Illustrating the History of Belgium.* Vol. 1: *From Prehistoric Times to 1830.* Memo from Belgium, no. 178. Brussels: Ministry of Foreign Affairs, External Trade and Cooperation in Development, 1978.

_____. *Documents Illustrating the History of Belgium.* Vol. 2: *Modern Belgium from 1830 to the Present Day.* Memo from Belgium, no. 180. Brussels: Ministry of Foreign Affairs, External Trade and Cooperation in Development, 1978.

Musée de la Vie Wallonne—Documents. Liège: Musée de la Vie Wallonne, 1914.

Unpublished Documents and Statements

AGLOP. "Le citoyen belge dans le système politique." A data set in the Belgian Archives for the Social Sciences, Louvain-la-Neuve, 1975.

"Appel aux Wallons." Petition presented at a press conference publicized under the title "Appel aux Wallons." Liège, Maison de la Presse, December 9, 1985.

Baere, M. de. "L'application de la loi du 14 juillet 1932 dans les établissements d'enseignement moyen et son influence sur les connaissances linguistiques des élèves." Centre de Recherche pour la Solution Nationale des Problèmes Sociaux, Politiques et Juridiques en Régions Wallonnes et Flamandes. Bibliothèque Royale Albert 1er, Brussels, doc. no. 18.

Dufrasne, M. "Communication concernant les recensements linguistiques." Centre de Recherche pour la Solution Nationale des Problèmes Sociaux, Politiques et Juridiques en Régions Wallonnes et Flamandes. Bibliothèque Royale Albert 1er, Brussels, doc. no. 137.

Maps and Atlases of Primary Value

Atlas de la Wallonie. 5 vols. Namur: Société de Développement Régional pour la Wallonie, 1980.

"La Belgique sous la domination française, 1794-1814: Les départements belges et les départements voisins au 1er janvier 1812." Collection of the Bibliothèque Royale Albert 1er, Brussels. Brussels: G. Van Oest, 1812.

Cambier, Erasme. "Chemin de fer de Tournai à Audenarde." Brussels: Etablissement Géographique de Bruxelles, 1865.

"Carte des chemins de fer concédés à la Compagnie Hainaut et Flandres." Brussels: Etablissement Géographique de Bruxelles, 1858.

"Chemin de fer projété de Tournay à St. Ghislain." Brussels: Etablissement Géographique de Bruxelles, 1850.

"Complément de la fusion des intérêts entre les provinces wallonnes et les Flandres—réseau des chemins de fer de l'Ouest Belge ou ligne de Braine-le-Compte à Courtrai avec embranchements sur Quenart et sur Gand." Brussels: Etablissement Géographique de Bruxelles, 1861.

Dainville, Françoise de, and Tuland, Jean. *Atlas administratif de l'empire français*. After the atlas produced by order of the Duc de Feltre in 1812. Paris: Librairie Minard, 1973.

"De Echte Taalgrens." Ronse: Kr.-Vl. Taalgrensaktiegroet Centraal Bureel, 1932. In *Varia 1-25*, 11, bound collection of the Biliothèque Royale Albert 1er, Brussels, n.d.

De Geyter, G. "België: Fusies van Gemeenten." Lier: J. Van In N.V., 1976.

Fris, V. "Kaart der Taalgrens in België en Noord-Frankrijk." Ghent: Geirnaert-Vandesteene, 1905.

Hochsteyn, Lucien. "Carte des communes du point de vue de la situation des langues nationales." Brussels: Institut Cartographique Militaire, 1907.

"Jonction des trois bassins houillers du Hainaut et des Flandres—chemin de fer de Braine-le-Compte à Gand." Brussels: Etablissement Géographique de Bruxelles, 1850.

"Projet de chemin de fer de Jemmapes à Nieuport—carte générale du tracé." Brussels: Etablissement Géographique de Bruxelles, 1845.

Newspapers and Journals

Bulletin de la Société Liègeoise de Littérature Wallonne (1858)
La Défense Wallonne (1891-1893; 1913)
Financial Times (1986)
La Flandre Libérale (1847)
Het Laatste Nieuws (1984-1986)
La Libre Belgique (1984-1986)
The New York Times (1986-1987)
La Nouvelle Revue Wallonne (1949-1962)
Le Soir (1984-1986)
De Standaard (1984-1986)
Thuis Zijn (1979-1984)
Union des Femmes de Wallonie (1913)
Wallonie (1984-1986)
Wallonie—Bruxelles: Une Communauté/Deux Régions (1983-1986)

Contemporary Works of Primary Value

Centre d'Etudes et de Recherches Urbaines. *Rénovation urbaine en Wallonie et à Bruxelles*. Brussels: Fondation Roi Baudouin, 1980.

Cosny, A. *Au beau pays de Flandre*. Brussels: Grands Annuaires, 1902.

De Raet, Lodewijk. *Over Vlaamse Volkskracht: De Vervlaamsching der Hoogeschool van Gent*. Brussels: De Vlaamsche Boekhandel, 1906.

_____. *Een Economisch Programma voor de Vlaamsche Beweging*. Brussels: De Vlaamsche Boekhandel, 1906.

_____. *Vlaanderen's Economische Ontwikkeling*. Brussels: Standaard, n.d.

Dekan, Rik. *Qui est qui en Belgique francophone*. Brussels: Editions BRD, 1981.

_____. *Wie is Wie in Vlaanderen 1980*. Brussels: Groep Cegos Makrotest, 1980.

Destrée, Jules. *Wallons et Flamands: la querelle linguistique en Belgique*. Paris: Plon-Nourrit, 1923.

Le développement de la Wallonie et les nouvelles technologies. Liège: Fondation André Renard, 1982.

De Wachter, Leo. *Repertorium van de Vlaamse Gouwen en Gemeenten: Algemeen Gedeelte en Gewesten*. Antwerp: De Sikkel, 1942.

Didaskalos [pseudonym]. *De 34 Open Brieven Geschreven door Didaskalos aan M. Helleputte, Minister van Spoorwegen, Posterijnen, Telegrafen en Telefonen en Verschenen in 'De Standaard' van 9 Mei 1907 tot April 1909*. Antwerp: De Standaard, 1909.

Duvieusart, Jean. *Wallonie 1938*. Liège: Editions de l'Action Wallonne, 1938.

Fonteyn, Guido; Van Mechelen, Loes; Van den Ende, Eliane; and Van De Vijver, Herman. *De Andere Belgen*. Brussels: BRT Open School, 1984.

Hasquin, Hervé, ed. *Communes de Belgique: dictionnaire d'histoire et de géographie administrative. Wallonie.* 2 vols. Brussels: La Renaissance du Livre, 1980.

_____. *Gemeenten van België: Geschiedkundig en Administratief Geografisch Woordenboek. Vlaanderen.* 2 vols. Brussels: La Renaissance du Livre, 1980.

Le Jeune, Rita, and Stiennan, Jacques, eds. *La Wallonie: le pays et les hommes.* 4 vols. Brussels: La Renaissance du Livre, 1977-1981.

Legros, Elisée. *A la recherche de nos origines wallonnes.* Liège: Editions des Forces Nouvelles, 1945.

De Ontwikkeling van de Vlaamse Economie in Internationaal Perspectief. Brussels: Bewestelijke Economische Raad voor Vlaanderen, n.d.

Van Aelbroeck, J. -L. *L'Agriculture pratique de la Flandre.* Paris: Madame Huzard, 1830.

Van Cauwelaert, Frans, et al. *Vlaanderen door de Eeuwen Heen.* Amsterdam: Elsevier, 1912.

Van Naelten, M. *Systeemtheoretische Verdenning van de Stedelijkheid in Vlaanderen.* 2 vols. Antwerp: Standaard Wetenschappelijke Uitgeverij, 1961.

van Overstraeten, Toon. *Op de Barrikaden: Het Verhaal van de Vlaamse Natie in Wording.* Tielt: Lannoo, 1984.

Vlaamsch-Nationalistisch Liederboek. Vilvoorde: Pieter Céonen, n.d.

Vlaanderen: Een Geografisch Portret. Antwerp: Gid, 1983.

Vlaanderen ons Vaderland. Antwerp: De Nederlandern, 1980.

Waarheen met België?/Que pourrait devenir la Belgique? Brussels: Center for the Study of Political Institutions, 1984.

Witte, Els., ed. *Geschiedenis van Vlaanderen.* Brussels: La Renaissance du Livre, 1983.

Selected Secondary Sources

General Published Works

Agnew, John A. "Structural and Dialectical Theories of Political Regionalism." *Political Studies from Spatial Perspectives.* Edited by Alan D. Burnett and Peter J. Taylor. New York: John Wiley & Sons, 1981. Pp. 275-289.

Akzin, Benjamin. *State and Nation.* London: Hutchinson University Library, 1964.

Alcock, Anthony E.; Taylor, Brian K.; and Welton, John M., eds. *The Future of Cultural Minorities.* New York: St. Martin's Press, 1979.

Alverson, Hoyt S. "The Roots of Time: A Comment on Utilitarian and Primordial Sentiments in Ethnic Identification." *Ethnic Autonomy: Comparative Dynamics, the Americas, Europe, and the Developing World.* Edited by Raymond L. Hall. New York: Pergamon Press, 1979. Pp. 13-17.

Ardrey, Robert. *The Territorial Imperative.* New York: Atheneum, 1966.

Atiyah, Patrick S. *Law and Modern Society.* Oxford: Oxford University Press, 1983.

Barry, Brian. "Reflections on Conflict." *Sociology,* 6 (1972), pp. 443-449.

Barth, Frederik, ed. *Ethnic Groups and Boundaries: The Social Organization of Cultural Difference.* Boston: Little, Brown & Co., 1969.

Barton, John H.; Gibbs, James Lowell, Jr.; Li, Victor Ha; and Merryman, John Henry. *Law in Radically Different Cultures.* St. Paul: West Publishing Co., 1983.

Bell, Daniel. "Ethnicity and Social Change." *Ethnicity: Theory and Experience.* Edited by Nathan Glazer and Daniel P. Moynihan. Cambridge, Massachusetts: Harvard University Press, 1975. Pp. 141-176.

Bell, Wendell, and Freeman, Walter, eds. *Ethnicity and Nation-Building: Comparative, International, and Historical Perspectives.* Beverly Hills: Sage, 1974.

Boeke, J. H. *Economics and Economic Policy of Dual Societies.* New York: Institute of Pacific Relations, 1953.

Cavin, Jean-François. *Territorialité, nationalité et droits politiques.* Lausanne: Held, 1971.

Clark, Gordon L., and Dear, Michael. *State Apparatus: Structures and Language of Legitimacy.* Boston: Allen & Unwin, 1984.

Clout, Hugh. *Regional Variations in the European Community.* Cambridge, England: Cambridge University Press, 1986.

Cobban, Alfred. *The Nation State and National Self-Determination.* New York: Thomas Y. Crowell, 1969.

Connelly, Brian, ed. "Political Integration in Multinational States." Entire issue of *Journal of International Affairs*, 1 (1973).

Connor, Walker. "Ethnic Nationalism as a Political Force." *World Affairs*, 133 (1970), pp. 91-97.

_____. "Nation-Building or Nation-Destroying?" *World Politics*, 24, 3 (April 1972), pp. 319-355.

_____. "Ethnonationalism in the First World: The Present in Historical Perspective." *Ethnic Conflict in the Western World.* Edited by Milton J. Esman. Ithaca: Cornell University Press, 1977. Pp. 19-45.

_____. "A Nation is a Nation, is a State, is an Ethnic Group, is a . . ." *Ethnic and Racial Studies*, 1, 4 (October 1978), pp. 377- 400.

_____. *The National Question in Marxist-Leninist Theory and Strategy.* Princeton: Princeton University Press, 1984.

Deutsch, Karl W. *Nationalism and Social Communication: An Inquiry into the Foundations of Nationality.* Cambridge, Massachusetts: Technology Press of M.I.T., 1953.

Deutsch, Karl W, and Foltz, William J., eds. *Nation-Building.* New York: Atherton, 1963.

DeVos, George. "Ethnic Pluralism: Conflict and Accommodation." *Ethnic Identity: Cultural Continuities and Change.* Edited by George DeVos and Lola Romanucci-Ross. Palo Alto, California: Mayfield Press, 1975. Pp. 5-41.

Dikshit, Ramesh D. *The Political Geography of Federalism.* New Delhi: Macmillan of India, 1975.

Dominion, Leon. *The Frontiers of Language and Nationality in Europe.* New York: American Geographical Society of New York and Henry Holt and Company, 1917.

Douglas, J. Neville H. "Political Integration and Division in Plural Societies: Problems of Recognition, Measurement and Salience." *Pluralism and Political Geography: People, Territory and State.* Edited by Nurit Kliot and Stanley Waterman. London: Croom Helm, 1983. Pp. 47-68.

Duchacek, Ivo D. *Comparative Federalism: The Territorial Dimension of Politics.* New York: Holt, Rinehart & Winston, 1970.

_____, ed. "Federalism and Ethnicity." *Publius*, 7, 4 (1977), entire issue.

Ehrlich, Stanislaw, and Wootton, Graham, eds. *Three Faces of Pluralism.* Westmead: Gower, 1980.

Eisenstadt, Samuel N., and Rokkan, Stein, eds. *Building States and Nations.* 2 vols. Beverly Hills: Sage, 1973.

Enloe, Cynthia H. *Ethnic Conflict and Political Development.* Boston: Little, Brown & Co., 1973.

Esman, Milton J., ed. *Ethnic Conflict in the Western World.* Ithaca: Cornell University Press, 1977.

_____. "Perspectives on Ethnic Conflict in Industrial Societies." *Ethnic Conflict in the Western World.* Edited by Milton J. Esman. Ithaca: Cornell University Press, 1977. Pp. 371-390.

Evenden, L. J., and Cunningham, F. F., eds. *Cultural Discord in the Modern World.* British Columbia Geographical Series, no. 20, 1973.

Fairchild, Henry Pratt. *Dictionary of Sociology.* New York: Philosophical Library, 1944.

Foster, Charles R., ed. *Nations Without a State: Ethnic Minorities in Western Europe.* New York: Praeger, 1980.

Furnivall, J. S. *Colonial Policy and Practice.* Cambridge, England: Cambridge University Press, 1948.

Gerth, Hans, and Mills, C. W. *Character and Social Structure.* New York: Harcourt, Brace and World, 1953.

Giddens, Anthony. *The Constitution of Society.* Berkeley: University of California Press, 1984.

Giner, Salvador, and Archer, Margaret S., eds. *Contemporary Europe: Social Structures and Cultural Patterns.* London: Weidenfeld and Nicholson, 1978.

Ginsburg, Norton S., and Roberts, Chester F., Jr. *Malaya.* Seattle: University of Washington, Press, 1958.

Gladdish, Kenneth R. "The Political Dynamics of Cultural Minorities." *The Future of Cultural Minorities.* Edited by Anthony E. Alcock, Brian K. Taylor, and John M. Welton. New York: St. Martin's Press, 1979. Pp. 161-176.

Glazer, Nathan, and Moynihan, Daniel P., eds. *Ethnicity: Theory and Experience.* Cambridge, Massachusetts: Harvard University Press, 1975.

Gold, John R. "Territoriality and Human Spatial Behavior." *Progress in Human Geography*, 6, 1 (March 1982), pp. 44-49.

Gordon, Milton M. "Toward a General Theory of Racial and Ethnic Group Relations." *Ethnicity: Theory and Experience.* Edited by Nathan Glazer and Daniel P. Moynihan. Cambridge, Massachusetts: Harvard University Press, 1975. Pp. 84-110.

Gottmann, Jean. "The Political Partitioning of Our World," and "Geography and International Relations." *World Politics*, 3, 2 (1951), pp. 153-173.

_____. *A Geography of Europe.* 3rd ed. New York: Holt, Rinehart & Winston, 1962.

_____. *The Significance of Territory.* Charlottesville: University of Virginia Press, 1973.

_____, ed. *Centre and Periphery: Spatial Variations in Politics.* London and Beverly Hills: Sage, 1980.

Gregory, Derek. *Ideology, Science and Human Geography.* New York: St. Martin's Press, 1978.

Hall, Raymond L., ed. *Ethnic Autonomy: Comparative Dynamics, the Americas, Europe, and the Developing World.* New York: Pergamon Press, 1979.

Hartshorne, Richard. "The Functional Approach in Political Geography." *Annals of the Association of American Geographers*, 60, 2 (1950), pp.95-130.

Hayes, Carlton. *The Historical Evolution of Modern Nationalism.* New York: Macmillan Press, 1931.

Hechter, Michael. *Internal Colonialism: The Celtic Fringe in British National Development.* Berkeley: University of California Press, 1975.

Héraud, Guy. "Pour un droit linguistique comparé." *Revue Internationale de Droit Comparé*, 23, 2 (April-June 1971), pp. 309-330.

_____. *L'Europe des ethnies.* Paris: Presses d'Europe, 1974.

Hoffman, George W. "Regional Policies and Regional Consciousness in Europe's Multinational Societies." *Geoforum*, 8 (1977), pp. 121-129.

Horowitz, Donald L. "Ethnic Identity." *Ethnicity: Theory and Experience.* Edited by Nathan Glazer and Daniel P. Moynihan. Cambridge, Massachusetts: Harvard University Press, 1975. Pp. 111-140.

Isaacs, Harold R. *Idols of the Tribe: Group Identity, and Political Change.* New York: Harper & Row, 1975.

Isajiw, Wsevolod. "Definitions of Ethnicity." *Ethnicity*, 1 (July 1974), pp. 111-124.

Johnston, Ray E., ed. *The Politics of Division, Partition and Unification.* New York: Praeger, 1976.

Johnston, Ronald J. *Geography and the State.* London: Arnold, 1982.

Jones, Stephen. "A Unified Field Theory of Political Geography." *Annals of the Association of American Geographers*, 64, 2 (1954), pp. 111-123.

Kauppi, Mark V. "The Resurgence of Ethno-Nationalism and Perspectives on State-Society Relations." *Canadian Review of Studies in Nationalism*, 11, 1 (Spring 1984), pp. 119-132.

Kliot, Nurit, and Waterman, Stanley. *Pluralism and Political Geography: People, Territory and State.* London: Croom Helm, 1983.

Kloss, H. "Territorialprinzip, Bekenntnisprinzip, Verfügungsprinzip: Über die Möglichkeiten der Abgrenzung der Volklichen Zugehörigkeit." *Europa Ethnica*, 22 (1965), pp. 52-73.

Knight, David B. "Identity and Territory: Geographical Perspectives on Nationalism and Regionalism." *Annals of the Association of American Geographers*, 72, 4 (1982), pp. 514-531.

_____. "Canada in Crisis: The Power of Regionalisms." *Tension Areas in the World.* Edited by D. G. Bennett. Champaign, Illinois: Park Press, 1982. Pp. 254-279.

_____. "Geographical Perspectives on Self-Determination." *Political Geography: Recent Advances and Future Directions.* Edited by Peter Taylor and John House. London: Croom Helm, 1984. Pp. 168-190.

Kohn, Hans. *The Idea of Nationalism: A Study in Its Origins and Background.* New York: Macmillan, 1944.

Lambert, Richard D. Preface to "Ethnic Conflict in the World Today." *The Annals of the American Academy of Political and Social Sciences*, 433 (September 1977), pp. vii-viii.

Laqueur, Walter, and Mosse, George, eds. "Nationalism and Separatism." Entire issue of *The Journal of Contemporary History*, 6, 1 (1971).

Lijphart, Arend. *Democracy in Plural Societies: A Comparative Exploration.* New Haven: Yale University Press, 1977.

_____. "Political Theories and the Explanation of Ethnic Conflict in the Western World: Falsified Predictions and Plausible Postdictions." *Ethnic Conflict in the Western World.* Edited by Milton J. Esman. Ithaca: Cornell University Press, 1977. Pp. 46-64.

Lipson, Leon, and Wheeler, Stanton. *Law and the Social Sciences.* New York: Russell Sage Foundation, 1985.

Lorwin, Val R. "Segmented Pluralism: Ideological Cleavages and Political Cohesion in the Smaller European Democracies." *Comparative Politics*, 3, 2 (January 1971), pp. 141-175.

Luethy, Herbert. "Incidences politiques de la pluralité des langues en Suisse." *Schweizer Monatshefte*, 46 (August 1966), pp. 401-408.

MacIver, Donald N. "Conclusion: Ethnic Identity and the Modern State." *National Separatism.* Edited by Colin H. Williams. Vancouver: University of British Columbia Press, 1982. Pp. 299-307.

Mackay, William F. *Bilinguisme et contact des langues.* Paris: Klincksiesh, 1976.

McRae, Kenneth D. "The Principle of Territoriality and the Principle of Personality in Multilingual States." *Linguistics,* 158 (August 1975), pp. 33-54.

Malmberg, Torsten. *Human Territoriality.* New York: Mouton, 1980.

Markusen, Ann R. "Regions and Regionalism." *Regional Analysis and the New International Division of Labor: Applications of a Political Economy Approach.* Edited by Frank Moulaert and Patricia W. Salinas. Boston: Kluwer & Nijhoff, 1983. Pp. 33-55.

Marx, Karl. *The Revolutions of 1848: Political Writings.* Vol. 1. Edited and translated from the German by David Fernbach. London: Allen Lane in association with the New Left Review, 1973.

Massey, Doreen. "Regionalism: Some Current Issues." *Capital and Class,* 6 (Autumn 1978), pp. 106-125.

Mikesell, Marvin W. "The Myth of the Nation State." *Journal of Geography,* 82, 6 (November-December 1983), pp. 257-260.

_____. "Culture and Nationality." *Geography in Internationalizing the Undergraduate Curriculum.* Edited by Salvatore J. Natoli and Andrew R. Bond. Washington, D.C.: Association of American Geographers, 1985. Pp. 67-80.

Mitchinson, Rosalind, ed. *The Roots of Nationalism: Studies in Northern Europe.* Atlantic Highlands, New Jersey: Humanities Press, 1984.

Mughan, Anthony. "Modernization and Regional Relative Deprivation: Towards a Theory of Ethnic Conflict." *Decentralist Trends in Western Democracies.* Edited by Lawrence J. Sharpe. Beverly Hills: Sage, 1979. Pp. 279-312.

Murphy, Alexander B. "Partitioning as a Response to Cultural Conflict." *Geographical Perspectives,* 5/*Great Plains-Rocky Mountain Geographical Journal,* 13 (Spring 1985), pp. 53-59.

Nairn, Tom. *The Break-up of Britain*. London: New Left Books, 1977.

Nelde, Peter Hans. "Language 'Contact Universals' along the Germanic-Romantic Linguistic Border." *Journal of Multilingual and Multicultural Development*, 2, 2 (1981), pp. 117-126.

Nicholls, David. *Three Varieties of Pluralism*. New York: St. Martin's Press, 1974.

Nordlinger, Eric. *Conflict Regulation in Divided Societies*. Harvard Studies in International Affairs, no. 29. Cambridge, Massachusetts: Center for International Affairs, Harvard University, 1972.

O'Barr, William M. "The Study of Language and Politics." *Language and Politics*. Edited by William M. O'Barr and Jean F. O'Barr. The Hague: Mouton, 1976. Pp. 1-27.

Orridge, Andrew W. "Separatist and Autonomist Nationalisms: The Structure of Regional Loyalties in the Modern State." *National Separatism*. Edited by Colin H. Williams. Vancouver: University of British Columbia Press, 1982. Pp. 43-74.

_____. "Varieties of Nationalism." *The Nation-State: The Formation of Modern Politics*. Edited by Leonard Tivey. Oxford: Robertson, 1981. Pp. 39-58.

Orridge, Andrew W., and Williams, Colin H.. "Autonomist Nationalism: A Theoretical Framework for Spatial Variations in Its Genesis and Development." *Political Geography Quarterly*, 1, 1 (January 1982), pp. 19-39.

Paddison, Ronan. *The Fragmented State: The Political Geography of Power*. New York: St. Martin's, 1983.

Palley, Claire. *Constitutional Law and Minorities*. London: Minority Rights Group Report No. 36, 1979.

Parkin, Frank. *Marxism and Class Theory: A Bourgeois Critique*. London: Travistock, 1979.

Parvin, Manoucher, and Sommer, Maurie. "Dar Al'Islam: The Evolution of Muslim Territoriality and Its Implications for Conflict Resolution in the Middle East." *International Journal of Middle East Studies*, 11, 1 (February 1980), pp. 1-21.

Philip, Alan Butt. "European Nationalism in the Nineteenth and Twentieth Centuries." *The Roots of Nationalism: Studies in Northern Europe.* Edited by Rosalind Mitchinson. Atlantic Highlands, New Jersey: Humanities Press, 1984. Pp. 1-9.

Pounds, Norman J. G. "History and Geography: A Perspective on Partition." *Journal of International Affairs*, 18, 2 (1964), pp. 161-172.

_____. *Political Geography.* 2nd ed. New York: McGraw Hill, 1972.

Pred, Allan. "Place as Historically Contingent Process: Structuration and the Time-Geography of Becoming Places." *Annals of the Association of American Geographers*, 74, 2 (1984), pp. 279-297.

Ra'anan, Uri, ed. *Ethnic Resurgence in Modern Democratic States. A Multidisciplinary Approach to Human Resources and Conflict.* Elmsford, New York: Pergamon Press, 1980.

Ragin, Charles C. "Ethnic Political Mobilization: The Welsh Case." *American Sociological Review*, 44, 4 (August 1979), pp. 619-634.

Rawkins, Phillip. "Nationalist Movements within the Advanced Nationalist State: The Significance of Culture." *Canadian Review of Studies in Nationalism*, 10, 2 (Fall 1983), pp. 221-233.

Riggs, Fred W. "What is Ethnic? What is National? Let's Turn the Tables." *Canadian Review of Studies in Nationalism*, 13, 1 (Spring 1986), pp. 111-123.

_____. *Ethnicity, INTERCOCTA Glossary, Concepts and Terms used in Ethnicity Research.* International Social Science Council Committee on Conceptual and Terminological Analysis. Honolulu: Department of Political Science, University of Hawaii, n.d.

Rokkan, Stein, and Urwin, Derek W., eds. *The Politics of Regional Identity: Studies in European Regionalism.* London: Sage, 1982.

Rudolph, Joseph R., Jr. "Ethnoregionalism in Contemporary Western Europe: The Potential for Political Accommodation." *Canadian Review of Studies in Nationalism*, 8, 2 (Fall 1981), pp. 323-341.

Sack, Robert D. "Territorial Bases of Power." *Political Studies from Spatial Perspectives.* Edited by Alan D. Burnett and Peter J. Taylor. New York: John Wiley & Sons, 1981. Pp. 53-71.

Sack, Robert D. "Human Territoriality: A Theory." *Annals of the Association of American Geographers*, 73, 1 (1983), pp. 55-74.

_____. *Human Territoriality: Its Theory and History*. Cambridge Studies in Historical Geography. Cambridge, England: Cambridge University Press, 1986.

Segal, Bernard E. "Ethnicity: Where the Present is the Past." *Ethnic Autonomy: Comparative Dynamics, the Americas, Europe, and the Developing World*. Edited by Raymond L. Hall. New York: Pergamon Press, 1979. Pp. 7-12.

Seymour, Charles. *Geography, Justice, and Politics at the Paris Conference of 1919*. New York: American Geographical Society, 1951.

Sharpe, Lawrence J., ed. *Decentralist Trends in Western Democracies*. Beverly Hills: Sage, 1979.

_____. "Decentralist Trends in Western Democracies: A First Appraisal." *Decentralist Trends in Western Democracies*. Edited by Lawrence J. Sharpe. Beverly Hills: Sage, 1979. Pp. 9-90.

Smith, Anthony D. *The Ethnic Revival in the Modern World*. Cambridge, England: Cambridge University Press, 1981.

_____. "Nationalism, Ethnic Separatism and the Intelligentsia." *National Separatism*. Edited by Colin H. Williams. Vancouver: University of British Columbia Press, 1982. Pp. 17-41.

_____. "Ethnic Identity and World Order." *Millennium: Journal of International Studies*, 12 (1982), pp.149-161.

Smith, M. G. "Social and Cultural Pluralism." *Annals of the New York Academy of Sciences*, 83, 5 (1960), pp. 786-795.

Snyder, Louis L. "Nationalism and the Territorial Imperative." *Canadian Review of Studies in Nationalism*, 3, 1 (Fall 1975), pp. 1-21.

_____. "Nationalism and the Flawed Concept of Ethnicity." *Canadian Review of Studies in Nationalism*, 10, 2 (Fall 1983), pp. 253-265.

Soja, Edward W. *The Political Organization of Space*. Commission on College Geography Resource Paper No. 8. Washington, D.C.: Association of American Geographers, 1971.

Soja, Edward W. "Regions in Context: Spatiality, Periodicity, and the Historical Geography of the Regional Question." *Environment and Planning D: Society and Space*, 3 (1985), pp. 175-190.

Stephens, Meic. *Linguistic Minorities in Western Europe*. Llandysul, Wales: Gomer Press, 1978.

Symmons-Symonolewicz, Konstantin. "The Concept of Nationhood: Toward a Theoretical Clarification." *Canadian Review of Studies in Nationalism*, 12, 2 (Fall 1985), pp. 215-222.

Tivey, Leonard. "States, Nations and Economies." *The Nation-State: The Formation of Modern Politics*. Edited by Leonard Tivey. Oxford: Robertson, 1981.

Van den Berghe, Pierre L. "Ethnic Pluralism in Industrial Societies: A Special Case." *Ethnicity*, 3 (1976), pp. 242-255.

Van Dyke, Vernon. "The Individual, the State and Ethnic Communities in Political Theory." *World Politics*, 29, 3 (April 1977), pp. 343-369.

_____. "Human Rights and the Rights of Groups." *American Journal of Political Science*, 18 (November 1974), pp. 725-741.

Wallerstein, Immanuel. "Ethnicity and National Integration in West Africa." *Cahiers d'Etudes Africaines*, I (1960), pp. 129-139.

Weinberg, Lee S., and Weinberg, Judith W. *Law and Society*. Washington, D.C.: University Press of America, Inc., 1980.

Whebell, C. F. J. "A Model of Territorial Separatism." *Proceedings of the Association of American Geographers*, 5 (1973), pp. 295-298.

Williams, Colin H. "The Territorial Dimension in Language Planning." *Language Problems and Language Planning*, 5, 1 (1981), pp. 57-73.

_____, ed. *National Separatism*. Vancouver: University of British Columbia Press, 1982.

_____. "Conceived in Bondage—Called unto Liberty: Reflections on Nationalism." *Progress in Human Geography*, 9, 3 (September 1985), pp. 331-355.

_____. "Perspectives on Minority Nationalism in Europe." *Nationalism in the Modern World*. Edited by John M. Walters. London: Croom Helm, 1986.

Williams, Colin H., and Smith, Anthony D.. "The National Construction of Social Space." *Progress in Human Geography*, 7,4 (1983) pp. 502-518.

Wirsing, Robert G., ed. *Protection of Ethnic Minorities: Comparative Perspectives*. New York: Pergamon Press, 1981.

_____. "Dimensions of Minority Protection." *Protection of Ethnic Minorities: Comparative Perspectives*. Edited by Robert G. Wirsing. New York: Pergamon Press, 1981. Pp. 3-17.

Wirth, Louis. "The Protection of Minority Groups." *The Science of Man in the World Crisis*. Edited by Ralph R. Linton. New York: Columbia University Press, 1945. Pp. 347-372.

Wolf, Ken. "Ethnic Nationalism: An Analysis and Defense." *Canadian Review of Studies in Nationalism*, 13, 1 (Spring 1986), pp. 99-109.

Young, Crawford. *The Politics of Cultural Pluralism*. Madison: University of Wisconsin Press, 1979.

Published Works on Belgium

André, Robert. "La dualité démographique de la Belgique." *Cahiers de Géographie du Québec*, 13, 30 (1969), pp. 321-332.

_____. "Eléments d'une politique démographique wallonne." *Wallonie* 74, 5 (1974), pp. 319-323.

_____. "La dualité démographique de la Belgique." *L'Ethnie Française*, 1ère partie, 3 (1981), pp. 133-154; 2ème partie, 4 (1981), pp. 221-254.

_____. "Migrations régionales en Belgique, 1962-1970. *Population et Famille*, 52 (1981), pp. 31-62.

André, Robert; Gossiaux, A-M; Pereira-Roque, J.; and Richez-Ruelens, E. "Evolution démographique et population active, le cas de la Wallonnie." *Wallonnie* 79, 5 (1979), pp. 497-530.

Arango, E. Ramon. *Leopold III and the Belgian Royal Question*. Baltimore: Johns Hopkins University Press, 1961.

Atlas de Belgique. Brussels: Comité National de Géographie, Presses de l'Institut Géographique Militaire, 1950-1972.

Atlas of Belgium. Brussels: Belgian Information and Documentation Institute, 1985.

"Autour du manifeste Schreurs-Couvreur." *La Nouvelle Revue Wallonne,* 5, 3 (April 1953), pp. 192-193.

Baetens Beardsmore, Hugo. "Bilingualism in Belgium." *Journal of Multilingual and Multicultural Development,* 2 (1980), pp. 145-154.

_____. "Linguistic Accommodation in Belgium." *Brussels Pre-Prints in Linguistics,* distributed by the linguistics circle of the Vrije Universiteit Brussel and the Université Libre de Bruxelles, 5 (March 1981), pp. 1-21.

Baeyens, Herman. "The Development of the Brussels Agglomeration." *Delta,* 6, 4 (Winter 1963-1964), pp. 79-89.

Basse, Maurits. *De Vlaamsche Beweging van 1905 tot 1930.* 2 vols. Ghent: Van Rysselberghe & Rombaut, 1930-1933.

Bauwens, Léon. *Régime linguistique de l'enseignement primaire et de l'enseignement moyen.* Brussels: Edition Universelle, 1933.

Becquet, Charles. "Interaction des ethnies dans la Belgique contemporaine." *Journal de la Société de Statistique de Paris,* 104, 4-6 (April-June 1963), pp. 104-131.

_____. *Le différend Wallo-Flamand.* 2 vols. Charleroi: Institut Jules Destrée, 1972-1977.

"Le bilinguisme de la Flandre." *Le Mouvement Géographique,* 35 (December 10, 1922), pp. 685-687.

Blauwkuip, Floris. *De Taalbesluiten van Koning Willem I.* Amsterdam: De Bussy, 1920.

Bologne, M. *Notre passé wallon,* Charleroi: Institut Jules Destrée, 1973.

Brassine, J. "La régionalisation: la loi du 1er août 1974 et sa mise en oeuvre, I-II." *Courrier Hebdomadaire du C.R.I.S.P.,* 665; 667/668 (December 12, 1974; January 10, 1975).

_____. "La réforme de l'état: Phase immédiate et phase transitoire." *Courrier Hebdomadaire du C.R.I.S.P.,* 857/858 (October 31, 1979).

Brusselse Randgemeenten: Een Onderzoek naar de Residentiele en Taalkundige Ontwikkeling door de Studiegroep 'Mens en Ruimte.' Antwerp: Kultuurraad van Vlaanderen, 1964.

Centre de Recherche et d'Information Socio-Politiques. "Les partis politiques non traditionnels." *Courrier Hebdomadaire du C.R.I.S.P.*, 101 (March 24, 1961).

_____. "Les élections législatives du 26 mars 1961." *Courrier Hebdomadaire du C.R.I.S.P.*, 104 (March 31 and April 7, 1961).

_____. "Tableau synthétique des projets de fédéralisme de 1931 à nos jours." *Courrier Hebdomadaire du C.R.I.S.P.*, 129 (November 14, 1961).

_____. "Le 'Manifeste des 29' et ses répercussions sur les structures politiques de la région bruxelloise I-III." *Courrier Hebdomadaire du C.R.I.S.P.*, 444/445; 448/449; 450 (May 5, 1969; June 20, 1969; June 27, 1969).

_____. "Les projets communautaires du gouvernement Eyskens-Merlot (1968) Eyskens-Cools (1969)." *Courrier Hebdomadaire du C.R.I.S.P.*, 451 (September 5, 1969).

_____. "L'évolution linguistique et politique du Brabant." *Courrier Hebdomadaire du C.R.I.S.P.*, 466/467 (January 16, 1970).

_____. "Du dialogue communautaire de l'hiver 1976-77 au pacte communautaire de mai 1977, I-III." *Courrier Hebdomadaire du C.R.I.S.P.*, 767; 772; 783/784 (June 6, 1977; September 9, 1977; December 16, 1977).

_____. "Le problème des Fourons de 1962 à nos jours." *Courrier Hebdomadaire du C.R.I.S.P.*, 859 (November 23, 1979).

_____. "Communautés et régions en Belgique." *Dossier Pédagogique du C.R.I.S.P.* (September 1984).

Ceuleers, Jan. "De Staatshervorming van 1980 als Niet-Oplossing." *Res Publica*, 26, 3 (1984), pp. 293-301.

Claes, Lode. "The Process of Federalization in Belgium." *Delta*, 6, 4 (Winter 1963-1964), pp. 43-52.

_____. "Le mouvement flamand entre le politique, l'économique et le culturel." *Res Publica*, 15, 2 (1973), pp. 219-236.

Claes, Lode. "Le mouvement flamand et la réforme de l'état." *La réforme de l'état*. Brussels: Centre d'Etudes pour la Réforme de l'Etat, 1977.

Claeys, Paul H. "Political Pluralism and Linguistic Cleavage: The Belgian Case." *Three Faces of Pluralism: Political, Ethnic and Religious*. Edited by Stanislaw Ehrlich and Graham Wootton. Westmead: Gower, 1980. Pp. 169-189.

Claval, Paul. "Géographie politique et aménagement du territoire en Belgique." *Revue Géographique de l'Est*, 2 (1962), pp. 167-170.

Clercq, M. de, and Naert, F. "Regional Economic Disparities in Belgium and Regional, National and Supra-National Policies: An Analysis of Collective Failure." *Planning Outlook*, 28, 1 (1985), pp. 14-20.

Cliquet, Robert L. "On the Differential Population Development of the Flemings and the Walloons and Its Influence on Flemish-Walloon Relations." *Homo*, 11, 1-2 (1960), pp. 67-88.

Clough, Shepard B. *A History of the Flemish Movement in Belgium: A Study in Nationalism*. New York: R. H. Smith, 1930.

Cocatre-Zilgien, André. "Les origines de la querelle linguistique belge." *Revue de Psychologie des Peuples*, 24, 2 (1969), pp. 129-137.

Coppieters, Frans. *The Community Problem in Belgium*. Brussels: Belgian Information and Documentation Institute, 1974.

Coppieters, Maurits. "De Vlaamse Volksbeweging." *Kultuurleven*, (November 1958), pp. 665-673.

Coulon, Marion. *L'Autonomie culturelle en Belgique*. Brussels: Fondation Charles Plisnier, 1961.

Covell, Maureen. "Ethnic Conflict and Elite Bargaining: The Case of Belgium." *West European Politics*, 4, 3 (October 1981), pp. 197-218.

Daubie, Christian. "Le pacte culturel: de sa genèse à son application." *Res Publica*, 17, 2 (1975), pp. 171-200.

Dauzat, Albert. "Le déplacement des frontières linguistiques du français de 1806 à nos jours." *La Nature*, 55, 2 (December 15, 1927), pp. 529-535.

De Bruyne, Arthur. *Het Federalisme in Vlaanderen*. Schepdaal, Holland: Het Pennoen, 1962.

Dehousse, J. M., and Destrée, Jules. "Les institutions fédérales dans les projets des socialistes wallons." *Res Publica*, 5 (1963), pp. 21-36.

De Jonghe, A. *De Taalpolitiek van Koning Willem I de Zuidelijke Nederlanden, 1814-1830*. Brussels: Steenlandt, 1967.

_____. "Het Vraagstuk Brussel de Duitse Flamenpolitik, 1940-1944. *Taal en Sociale Integratie*, 4 (1981), pp. 405-443.

De Lannoy, W., and Declerck, Hugo. "Migraties naar het Brusselse Randgebied: Het Geval Dilbeek." *Taal en Sociale Integratie*, 3 (1981), pp. 269-280.

De Lannoy, W., and en Rampelbergh, M. *Sociaal-Geografische Atlas van Brussel en Zijn Randgebied*. Antwerpen: Ed Sikkel, 1981.

Delattre, Louis. *Le pays wallon*. Brussels: J. Lebèque & Co, 1905.

Deleu, Jozef; Durnez, Gaston; de Schryver, Reginald; and Simons, Ludo, eds. *Encyclopedie van de Vlaamse Beweging*. Tielt: Lannoo, 1975.

Delruelle, Nicole. *La mobilité sociale en Belgique*. Brussels: Editions de l'Institut de Sociologie, 1970.

Delruelle, Nicole; Coenen, Jacques; and Maigray, Daniel. "Les problèmes qui préoccupent les Belges: février-mars 1966." *Revue de l'Institut de Sociologie* (1966), pp. 291-341.

Delruelle, Nicole, and Frognier, André Paul. "L'opinion publique et les problèmes communautaires." *Courrier Hebdomadaire du C.R.I.S.P.*, 880; 927/928; 966 (May 9, 1980; July 3, 1981; June 11, 1982).

Delvaux, Yves. "Une certaine image de la Flandre ... et nous." *La Revue Nouvelle*, 38, 10 (1982), pp. 320-326.

De Metsenaere, Machteld. "Taalkaart van Belgie (1846-1866-1880)." *Taal en Sociale Integratie*, 2 (1979), pp. 41-75.

Demoulin, Robert. *La révolution de 1830*. Brussels: La Renaissance du Livre, 1950.

Deneckere, Marcel. *Histoire de la langue française dans les Flandres 1770-1823*. Ghent: Romancia Gandensia, 1954.

De Nolf, Rigo. "Federalism in Belgium as a Constitutional Problem." *Res Publica*, 8, 3 (1968), pp. 383-406.

De Schryver, Reginald. "The Belgian Revolution and the Emergence of Belgium's Biculturalism." *Conflict and Coexistence in Belgium: The Dynamics of a Culturally Divided Society.* Edited by Arend Lijphart. Berkeley: Institute of International Studies, 1981. Pp. 13-33.

Desonay, Fernand. "Two Ways of Looking at Flanders: A Walloon View." *Delta*, 6, 4 (Winter 1963-1964), pp. 33-39.

De Vroede, Maurits. *The Flemish Movement in Belgium.* Antwerp: Kultuurraad voor Vlaanderen, 1975.

Dhondt, J. "Essai sur l'origine de la frontière linguistique." *L'Antiquité Classique*, 16 (1947), pp. 261-286.

_____. "Note sur l'origine de la frontière linguistique." *L'Antiquité Classique*, 21 (1952), pp. 107-122.

Dickson, Tim. "High Cost of Cultural Divide." *Financial Times*, June 13, 1986.

Driesen, Ivo, and Swalens, Guido. "Bevolkingsmigraties in Sint-Genesius-Rode (1945-1975)." *Taal en Sociale Integratie*, 1 (1978), pp.153-198.

Dufrasne, A. "Au sujet de la communication <<Interaction des problèmes linguistiques et economiques en Belgique>>." *Journal de la Société de Statistique de Paris*, 104, 7-9 (July-September 1963), pp. 180-181.

Dumont, M. E., and De Smet, L. *Aardrijkskundige Bibliografie van België Bibliographie géographique de la Belgique.* 4 vols. Ghent: Seminarie voor Menselijke Aardrijkskunde der Rijksuniversiteit, 1954-1956 (Bibliographica Belgica, 14-17).

_____. *Eerste Aanvulling/Premier supplément.* Ghent: Aardrijkskundig Instituut der Rijksuniversiteit, 1960 (Bibliographica Belgica, 48).

_____. *Tweede Aanvulling/Deuxième supplément.* Ghent: Aardrijkskundig Instituut der Rijksuniversiteit, 1965 (Bibliographica Belgica, 82).

_____. *Derde Aanvulling/Troisième supplément.* Ghent: Publikaties van het Seminarie voor Menselijke en Ekonomische Geographie 3, 1970 (Bibliographica Belgica, 113).

Dumont, M. E.; De Smet, L.; and W. *Vierde. Aanvulling/Quatrième Supplément.* Ghent: Publikaties van het Seminarie voor Menselijke en Ekonomische Geographie 14, 1978 (Bibliographica Belgica, 113).

_____. *Vijfde Aanvulling/Cinquième Supplément.* Ghent: Publikaties van het Seminarie voor Menselijke en Ekonomische Geographie 15, 1980 (Bibliographica Belgica, 113).

Dunn, James A., Jr. "The Revision of the Constitution in Belgium: A Study in the Institutionalization of Ethnic Conflict." *Western Political Quarterly,* 27, 1 (March 1974), pp. 143-163.

DuRoy, Albert. *La guerre des Belges.* Paris: Seuil, 1968.

Elias, Hendrik Jozef. *Geschiedenis van de Vlaamse Gedachte, 1780-1914.* 4 vols. Antwerp: De Nederlandsche Boekhandel, 1963-1965.

_____. *Vijfentwintig Jaar Vlaamse Beweging, 1914-1939.* 4 vols. Antwerp: De Nederlandsche Boekhandel, 1969.

"Extension du mouvement wallon." *La Défense Wallonne,* 1, 8 (March 1, 1982).

Féaux, Valmy. *Cinq semaines de lutte sociale: la grève de l'hiver 1960-1961.* Brussels: Institut de Sociologie de l'Université Libre de Bruxelles, 1963.

Fitzmaurice, John. *The Politics of Belgium: Crisis and Compromise in a Plural Society.* London: C. Hurst & Company, 1983.

Fonteyn, Guido. *Les Wallons.* Brussels: Oyez, 1979.

Fox, Renée C. "Why Belgium?" *European Journal of Sociology,* 19, 2 (1978), pp. 205-228.

Fredericq, Paul. *Schets eener Geschiedenis der Vlaamsche Beweging.* Issued as Volumes II and III (III is in 2 parts) of *Vlaamsch België sedert 1830.* Ghent: Vuylsteke, 1906-1909.

George, Pierre, and Sevrin, Robert. *Belgique, Pays-Bas, Luxembourg.* Paris: Presses Universitaires de France, 1967.

Gérain, René. "La responsabilité politique des ministres devant les conseils culturels." *Res Publica,* 17, 1 (1975), pp. 31-52.

Gielen, Gerda, and Louckx, Freddy. "Sociologisch Onderzoek naar de Herkomst het Taalgedrag en het Schoolkeuzegedrag van Ouders met Kinderen in het Nederlandstalig Basisonderwijs van de Brussels Agglomeratie." *Taal en Sociale Integratie*, 7 (1984), pp. 161-208.

Gilissen, John. "La constitution belge de 1831: ses sources, son influence." *Res Publica*, 10, special supplement (1968), pp. 107-141.

Goffert, V. "La crise de Louvain, du 1er janvier au 31 mars 1968." *Res Publica*, 11 (1969), pp. 31-76.

Goris, Jan-Albert. *Belgium*. Berkeley: University of California Press, 1946.

Gysseling, Maurits. "Vlaanderen (Etymologie en Betekenisevolutie)." *Encyclopedie van de Vlaamse Beweging*. Edited by Jozef Deleu, Gaston Durnez, Reginald de Schryver, and Ludo Simons. Tielt: Lannoo, 1975. Pp. 1906-1912.

Goriely, Georges. "Frontière linguistique et destin de la Belgique." *L'Histoire des doctrines politiques contemporaines: essais sur le nationalisme*. Edited by Georges Goriely. Brussels: Presses Universitaires de Bruxelles, 1982-1983. Pp. 70-86.

Hamélius, Paul. *Histoire politique et littéraire du mouvement flamand au 19e siècle*. 2nd ed. Brussels: L'Eglantine, 1925.

Hamers, Josiane. "The Language Question in Belgium." *Language and Society*, 5 (Spring/Summer 1981), pp. 17-20.

Heisler, Martin O. "Institutionalizing Societal Cleavages in a Cooptive Polity: The Growing Importance of the Output Side in Belgium." *Politics in Europe: Structures and Processes in Some Postindustrial Democracies*. Edited by Martin O. Heisler. New York: McKay, 1974. Pp. 178-220.

Hendricks, Henri. "Le français dans les secteurs financier et industriel en Flandre." *La langue française dans les pays du Benelux: besoins et exigences*. Edited by Eddy Rosseel. Brussels: Association Internationale pour la Recherche et la Diffusion des Méthodes Audio-Visuelles et Structuro-Globales, 1982. Pp. 67-75.

Henrad, Maurice. *L'emploi de langues dans l'administration et dans les entreprises privées (loi du août 1963)*. Heule: Editions Administratives, 1964.

Henry, Albert. *Esquisse d'une histoire des mots "Wallon" et "Wallonie."* Brussels: La Renaissance du Livre, 1974.

_____. "Wallon et Wallonie." *La Wallonie: le pays et les hommes.* Vol. I. Edited by Rita Le Jeune and Jacques Stiennon. Brussels: La Renaissance du Livre, 1977. Pp. 67-76.

Hermans, Michel, and Verjans, Pierre. "Les origines de la querelle fouronaise." *Courrier Hebdomadaire du C.R.I.S.P.*, 1019 (December 2, 1983).

Herremans, Maurice-Pierre. *La question flamande.* Brussels: Librairie Meurice, 1948.

_____. *La Wallonie: ses griefs, ses aspirations.* Brussels: Editions Marie-Julienne, 1951.

_____. "Bref historique des tentatives de réforme du régime unitaire en Belgique." *Courrier Hebdomadaire du C.R.I.S.P.*, 135 (January 12, 1962).

Herremans, Maurice-Pierre, and Coppieters, Frans. *The Language Problem in Belgium.* Brussels: Belgian Information and Documentation Institute, 1967.

Huggett, Frank E. "Communal Problems in Belgium." *The World Today*, 22, 10 (October 1966), pp. 446-452.

_____. *Modern Belgium.* New York: Praeger, 1969.

Huyse, Luc. "The Language Conflict in Belgium: A Sociological Approach." *Sociological Contributions from Flanders.* Vol. 4. Deurne: Kluwer, 1975.

L'Institut Belge de Science Politique—Projet AGLOP-GLOPO. "Les citoyens belges et leur conception du monde politique." *Res Publica*, 17, 2 (1975), pp. 319-325.

INUSOP-UNIOP and Le Centre de Sociologie Générale, *Bruxelles et sa banlieue: opinions et attitudes des habitants à l'égard des problèmes politiques, linguistiques, sociaux, urbains et culturels. Enquête sociologique.* Brussels: Institut de Sociologie, Université Libre de Bruxelles, 1985.

Irving, R. E. M. *The Flemings and Walloons of Belgium.* London: Minority Rights Group Report, 1980.

"It's Hard Going: A Survey of Belgium." *The Economist*, February 22, 1986, special section, 18 pp.

Jennissen, Emile. *Le mouvement wallon*. Liège: Imprimerie La Meuse, 1913.

Jouret, B. *Définition spatiale du phénomène urbain bruxellois*. Brussels: Editions de l'Université de Bruxelles, 1972.

Kabugubugu, Amédée, and Nuttin, Joseph R.. "Changement d'attitude envers la Belgique chez les étudiants flamands." *Psychologica Belgica*, 11 (1970-1971), pp. 23-44.

Kane, Jean Ellen. "Flemish and Walloon Nationalism: Devolution of a Previously Unitary State." *Ethnic Resurgence in Modern Democratic States: A Multidisciplinary Approach to Human Resources and Conflict*. Edited by Uri Ra'anan. Elmsford, New York: Pergamon Press, 1980. Pp. 122-171.

Kelly, George A. "Belgium: New Nationalism in an Old World." *Comparative Politics*, 1, 3 (April 1969), pp. 343-365.

Kerkhofs, Jan. "Orientations dans le domaine de l'ethnique." *Univers des Belges*. Edited by Rudolph Rezsohazy and Jan Kerkhofs. Louvain-la-Neuve: CIACO, 1984.

Köhler, Ludwig von. *Die Staatsverwaltung der Besetzten Gebiete Belgiens*. New Haven: Yale University Press, 1927.

Kurth, Godefroid. *La frontière linguistique en Belgique et dans le nord de la France*. Brussels: Académie Royale des Sciences, des Lettres et des Beaux-Arts de Belgique, 1895-1898.

Lambert, André, and Sonnet, Anne. "Population, emploi et inactivité dans les trois régions linguistiques: quel avenir? Des réponses issues du modèle odyssée." *Reflets et Perspectives de la Vie Economique*, 18, 4-5 (1979), pp. 327-346.

Larmuseau, H.; Desmedt, F.; Lambrecht, M.; Damas, H.; and Wattelar, C. "Nouvelles perspectives de population (1976-2000) pour la Belgique, ses régions et ses arrondissements." *Bulletin de Statistique* 66, 3 (1980), pp. 200-252.

Laurent, A., and Declercq-Tijgat, A. "Les migrations internes définitives relatives aux agglomérations de Bruxelles, Liège, Charleroi, Verviers et Namur." *Population et Famille*, 45, 3 (1978), pp. 73-132.

Lefèvre, Jacques. "Dialect and Regional Identification in Belgium: The Case of Wallonia." *International Journal of the Sociology of Language*, 15 (1978), pp. 47-51.

_____. "Nationalisme linguistique et identification linguistique: le cas de Belgique." *International Journal of the Sociology of Language*, 20 (1979), pp. 37-58.

Lefèvre, Marguerite A. "Conditions de l'évolution de la Flandre, région géographico-historique." *Bulletin de la Société Belge d'Etudes Géographiques*, 36, 1 (1967), pp. 23-36.

Legros, Elisée. *La frontière des dialectes romans en Belgique.* Liège: Vaillant-Carmanne, 1948.

_____. "La frontière linguistique en Belgique." *Onomastica*, 2 (1948), pp. 9-16.

Leroy, R.; Godano, A.; and Sonnet, A. "La configuration spatiale de la crise de l'emploi." *Courrier Hebdomadaire du C.R.I.S.P.*, 1023/1024 (December 23, 1983).

Levy, Paul M. G. "La statistique des langues en Belgique." *Revue de l'Institut de Sociologie*, no. 3 (July-Sept. 1938), pp. 507-570.

_____. *La querelle du recensement.* Brussels: Bibliothèque de L'Institut Belge de Science Politique, serie 3, 1960.

_____. "La mort du recensement linguistique." *La Revue Nouvelle*, 36, 9 (1962), pp 145-154.

_____. "Quelques problèmes de statistique linguistique à la lumière de l'expérience belge." *Revue de l'Institut de Sociologie* (1964), 251-273.

_____. "Linguistic and Semantic Borders in Belgium." *International Journal of the Sociology of Language*, 15 (1978), pp. 9-19.

Lijphart, Arend, ed. *Conflict and Coexistence in Belgium: The Dynamics of a Culturally Divided Society.* Berkeley: Institute of International Studies, 1981.

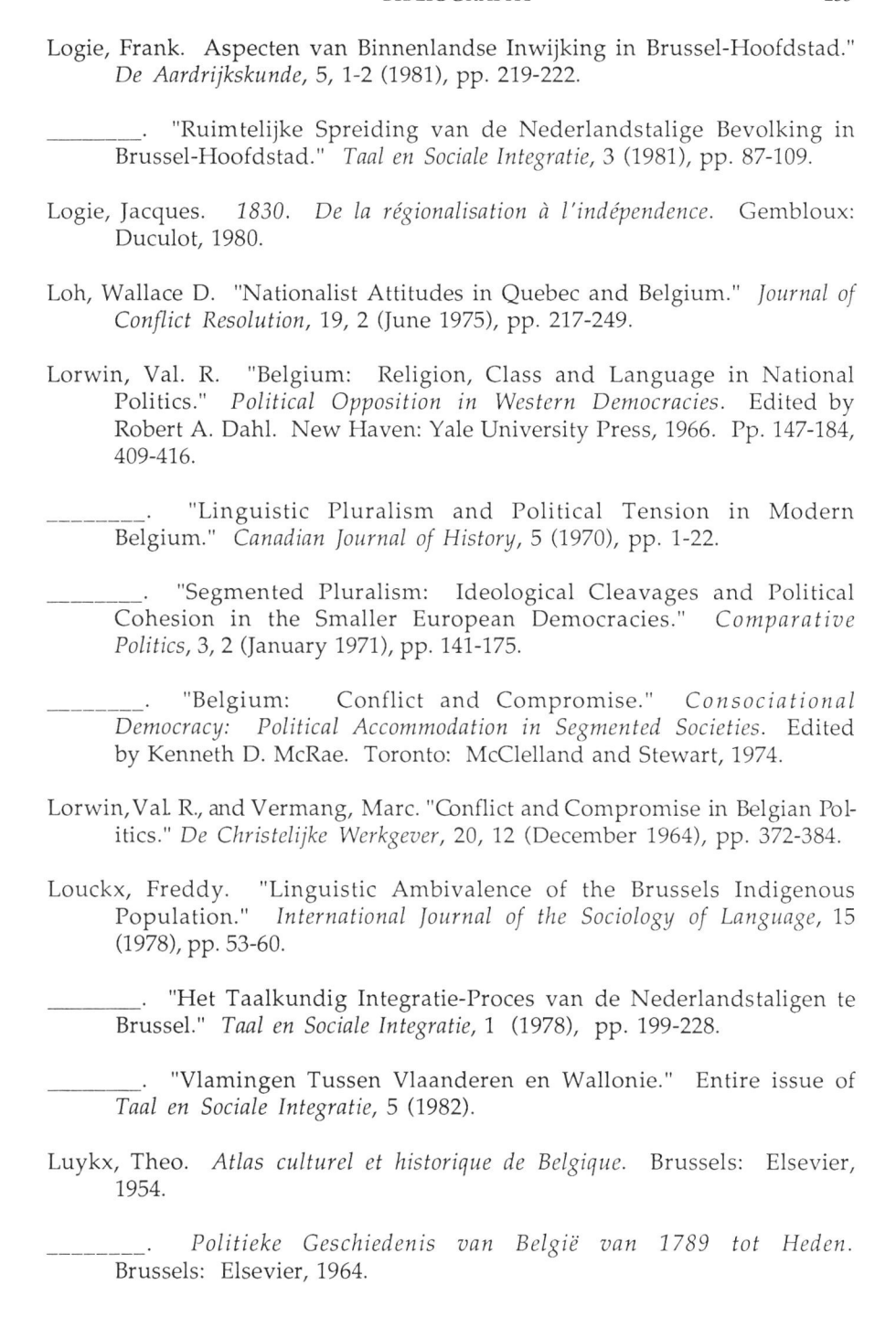

Logie, Frank. Aspecten van Binnenlandse Inwijking in Brussel-Hoofdstad." *De Aardrijkskunde*, 5, 1-2 (1981), pp. 219-222.

_____. "Ruimtelijke Spreiding van de Nederlandstalige Bevolking in Brussel-Hoofdstad." *Taal en Sociale Integratie*, 3 (1981), pp. 87-109.

Logie, Jacques. *1830. De la régionalisation à l'indépendence.* Gembloux: Duculot, 1980.

Loh, Wallace D. "Nationalist Attitudes in Quebec and Belgium." *Journal of Conflict Resolution*, 19, 2 (June 1975), pp. 217-249.

Lorwin, Val. R. "Belgium: Religion, Class and Language in National Politics." *Political Opposition in Western Democracies.* Edited by Robert A. Dahl. New Haven: Yale University Press, 1966. Pp. 147-184, 409-416.

_____. "Linguistic Pluralism and Political Tension in Modern Belgium." *Canadian Journal of History*, 5 (1970), pp. 1-22.

_____. "Segmented Pluralism: Ideological Cleavages and Political Cohesion in the Smaller European Democracies." *Comparative Politics*, 3, 2 (January 1971), pp. 141-175.

_____. "Belgium: Conflict and Compromise." *Consociational Democracy: Political Accommodation in Segmented Societies.* Edited by Kenneth D. McRae. Toronto: McClelland and Stewart, 1974.

Lorwin, Val. R., and Vermang, Marc. "Conflict and Compromise in Belgian Politics." *De Christelijke Werkgever*, 20, 12 (December 1964), pp. 372-384.

Louckx, Freddy. "Linguistic Ambivalence of the Brussels Indigenous Population." *International Journal of the Sociology of Language*, 15 (1978), pp. 53-60.

_____. "Het Taalkundig Integratie-Proces van de Nederlandstaligen te Brussel." *Taal en Sociale Integratie*, 1 (1978), pp. 199-228.

_____. "Vlamingen Tussen Vlaanderen en Wallonie." Entire issue of *Taal en Sociale Integratie*, 5 (1982).

Luykx, Theo. *Atlas culturel et historique de Belgique.* Brussels: Elsevier, 1954.

_____. *Politieke Geschiedenis van België van 1789 tot Heden.* Brussels: Elsevier, 1964.

Luykx, Theo, and Platel, Marc. *Politieke Geschiedenis van Belgie van 1944 tot 1985.* Antwerp: Kluwer Rechtswetenschappen, 1985.

Mabille, Xavier. "Les facteurs d'instabilité gouvernementale: décembre 1978-avril 1981." *Courrier Hebdomadaire du C.R.I.S.P.*, 916 (April 10, 1981).

Macar, Paul. "Un important problème wallon: la querelle linguistique en Belgique." *Revue Canadienne de Géographie,* 17, 3-4 (1963), pp. 137-143.

McRae, Kenneth D. *Conflict and Compromise in Multilingual Societies: Belgium.* Waterloo, Ontario: Wilfrid Laurier University Press, 1986.

Maes, Rudolf. *La décentralisation territoriale: situation et perspectives.* Report to the Ministry of the Interior and the Public Service. Brussels: INBEL, 1985.

Maeyer, A. de. "Wijzigingen van de Gemeentegrenzen in België." *La Géographie/De Aardrijkskunde,* 16, 4 (1964), pp. 272-284.

Mandel, Ernest. "The Dialectic of Class and Region in Belgium." *New Left Review,* 20 (Summer 1963), pp. 5-31.

A Manual of Belgium and the Adjoining Territories: Atlas. London: Naval Staff Intelligence Division, 1918.

Maroy, Pierre. "L'évolution de la législation linguistique belge." *Revue du Droit Public et de la Science Politique en France,* 82 (1969), pp. 449-501.

Monteyne, André. *De Brusselaars in een Stad die Anders is.* Tielt-Bussum: Lannoo, 1981.

Mughan, Anthony. "Modernization and Ethnic Conflict in Belgium." *Political Studies,* 27, 1 (March 1979), pp. 21-37.

Murphy, Alexander B. "Evolving Regionalism in Linguistically Divided Belgium." *Nationalism, Self-Determination, and Political Geography.* Edited by R.J. Johnston, David Knight, and Eleanor Kofman. London: Croom Helm, 1988. Pp. 135-150.

Nielsen, François. "The Flemish Movement in Belgium after World War II: A Dynamic Analysis." *American Sociological Review,* 45, 1 (February 1980), pp. 76-94.

Nuttin, J. M., Jr. "De Ontwikkeling van de Gezindheid Tegenover de Walen en het Persoonlijk Contact." *Tijdschrift voor Opvoedkunde*, 5 (1969), pp. 315-333.

Nuttin, Joseph. "Het Stereotiep Beeld van Walen, Vlamingen en Brusselaars: Hun Kijk op Zichzelf en op Elkaar." *Mededelingen van de Koninklijke Academie voor Wetenschappen, Letteren en Schone Kunsten*, Klasse der Letteren, 38, 2 (1976).

Obler, Jeffrey L. "Group Rights and the Linguistic Dispute in Brussels." *Taal en Sociale Integratie*, 3 (March 1981), pp. 39-59.

Outers, Lucien. *Le divorce belge.* Paris: Les Editions de Minuit, 1968.

Petri, Franz. *Germanische Volkserbe in Wallonien und Nord-Frankreich: Die Fränkische Landnahme in Frankreich und den Niederlanden und die Bildung der Westlichen Sprachgrenze.* 2 vols. Bonn: L. Röhrscheid, 1937.

Philippart, André. "Belgium: Language and Class Oppositions." *Government and Opposition*, 2, 1 (October 1966—January 1967), pp. 63-87.

Picard, H. "De Talentelling in en Rond Brussel." *Vlaamse Gids*, 39 (1955), pp. 5-22.

Pirenne, Henri. *Histoire de Belgique.* 7 vols. Brussels: Lamertin, 1902-1932.

Platel, Marc. "Bruxelles, aussi capitale des Flamands." Translated from the Dutch by Willy Devos. *Septentrion-Revue de Culture Néerlandaise*, 9, 2 (June 1980), pp. 77-81.

Plavsic, Vladimar S. "Les régions, les provinces et les communes en quete d'autonomie." *Res Publica*, 15, 5 (1973) pp. 915-946.

Polasky, Janet. "Liberalism and Biculturalism." *Conflict and Coexistence in Belgium: The Dynamics of a Culturally Divided Society.* Edited by Arend Lijphart. Berkeley: Institute of International Relations, 1981. Pp. 34-45.

Preudhomme, Claude. "Esquisse de la Belgique 'régionalisée.'" *Bulletin Trimestriel du Crédit Communal de Belgique*, 30, 116 (1976), pp. 91-106; 117 (1976), pp. 201-209; 118 (1976), pp. 217-240.

Quévit, Michel. *Les causes du déclin wallon.* Brussels: Editions Vie Ouvrière, 1978.

Quévit, Michael. *La Wallonie: l'indispensable autonomie.* Brussels: Entente, 1982.

Riley, Raymond. *Belgium.* Folkestone, England: Dawson, 1976.

Rosseel, Eddy. "Le français dans l'éducation en Flandre." *La langue française dans les pays du Benelux: besoins et exigences.* Edited by Eddy Rosseel. Brussels: Association Internationale pour la Recherche et la Diffusion des Méthodes Audio-Visuelles et Structuro-Globales, 1982. Pp. 76-89.

Rudolph, Joseph R., Jr. "Belgium: Controlling Separatist Tendencies in a Multinational State." *National Separatism.* Edited by Colin H. Williams. Vancouver: University of British Columbia Press, 1982. Pp. 263-297.

Ruys, Manu. *The Flemings: A People on the Move, A Nation in Being.* Translated from the Dutch by Henri Schoup. Tielt: Lannoo, 1973.

Schiltz, Hugo. *Uitdaging aan de Vlaamse Meerderheid.* Belgium: Soethoudt, 1985.

Schreurs, Fernand. *Les congrès de Rassemblement Wallon de 1890 à 1959.* Liège: Institut Jules Destrée, 1950.

_____. "A propos de l'agglomération bruxelloise." *La Nouvelle Revue Wallonne,* 7, 4 (3rd trimester 1955), pp. 233-240.

Seewald, Ulrich. "Der Flämisch-Wallonische Konflikt in Belgien." *Geographische Rundschau,* 22, 7 (July 1970), pp. 257-265.

Senelle, Robert. *The Political, Economic and Social Structures of Belgium.* Memo from Belgium, nos. 122-124. Brussels: Ministry of Foreign Affairs, External Trade and Cooperation in Development, 1970.

_____. *The Revision of the Constitution, 1967-1970.* Memo from Belgium, nos. 132-133. Brussels: Ministry of Foreign Affairs, External Trade and Cooperation in Development, 1971.

_____. *The Reform of the Belgian State.* 3 vols. Memo from Belgium, nos. 315, 319, and 326. Brussels: Ministry of Foreign Affairs, External Trade and Cooperation in Development, 1978-1980.

Servais, Paul. "Le sentiment national en Flandres et en Wallonie: approches socio-linguistiques." *Recherches Sociologiques,* 2 (1970), pp. 123-144.

Sevrin, Robert. *Géographie de la Belgique et des Pays-Bas.* Paris: Presses Universitaires de France, 1969.

Seyn, Eugene de. *Dictionnaire historique et géographique des communes belges.* 3rd ed. Turnhout: Brepols, 1950.

Solansky, Adolf. *German Administration in Belgium.* New York: Columbia Studies, 1928.

Sporck, José, and Christians, Charles. "L'organisation régionale de l'espace en Belgique et au Grand Duché de Luxembourg." *Revue Géographique de l'Est,* 16 (1976), pp. 121-150.

Sporck, José, and Goosens, Modest. "Le réseau urbain: les zones d'influence des villes et la hiérarchie urbaine." *La cité belge d'aujourd'hui: quel devenir?,* Bulletin Trimestriel du Crédit Communal de Belgique, no. 154. Brussels: Crédit Communal de Belgique, 1985. Pp. 192-197.

Stengers, Jean. "Belgian National Sentiments." *Conflict and Coexistence in Belgium: The Dynamics of a Culturally Divided Society.* Edited by Arend Lijphart. Berkeley: Institute of International Studies, 1981. Pp. 46-60.

_____. *La formation de la frontière linguistique en Belgique, ou de la légitimité de l'hypothèse historique.* Brussels: Collection Latomus, vol. 41, 1959.

Stephenson, Glenn V. "Cultural Regionalism and the Unitary State Idea in Belgium." *The Geographical Review,* 62, 4 (October 1972), pp. 501-523.

Stexhe, Paul de. *La révision de la constitution belge: 1968-1971.* Namur: Société d'Etudes Morales, Sociales et Juridiques, 1972.

Swing, Elizabeth. *Bilingualism and Linguistic Segregation in the Schools of Brussels.* Québec: Centre International de Recherche sur le Bilinguisme, 1980.

Swyngedouw, Eric. *Contradictions between Economic and Physical Planning in Belgium.* Working Paper No. 3. Villeneuve d'Ascq: Johns Hopkins European Center for Regional Planning and Research, 1985.

ter Hoeven, P. J. Augustinus. "The Social Bases of Flemish Nationalism." *International Journal of the Sociology of Language,* 15 (1978), pp. 21-32.

Tilsey, M. "Het Vestigingspatroon van de Financiele Instellingen in Belgie." *Belgische Vereniging voor Aardrijkskundige Studies*, 48, 1 (1979), pp. 255-296.

Tindemans, Léo. *L'autonomie culturelle*. Brussels: Van Ruys, 1971.

_____. "Discussion: rapport introductif sur 'Bruxelles et le fédéralisme.'" *Res Publica*, 13, 3-4 (1971), pp. 430-436.

Tulp, Stella. "Reklame en Tweetaligheid een Onderzoek naar de Geografische Verspreiding van Franstalige en Nederlandstalige Affiches in Brussel." *Taal en Sociale Integratie*, 1 (1978), pp. 261-288.

Uhlig, Harald. "Belgien: Entwicklung, Struktur und Raumliche Glederung Seiner Bevölkerung." *Geographisches Taschenbuch. Jahrweiser zur Deutschen Landeskunde*, 1958-1959. Wiesbaden: Franz Steiner Verlag, 1958. Pp. 373-389.

Valkhoff, W. *Geschiedenis en Actualiteit der Frans-Nederlandse Taalgrens*. Amsterdam: J. M. Meulenhoff, 1950.

Van de Craen, Piet, and Langenakens, Ann. "Verbale Strategieen bij Nederlandstaligen in Sint-Genesius-Rode." *Taal en Sociale Integratie*, 2 (1979), pp. 97-139.

Van den Berghe, Patrick, and Wils, Lod. "De Wet van 22 Februari 1908 op het Taalgebruik in Strafzaken, Bijzonder het Arrondissement Brussels." *Taal en Sociale Integratie*, 7 (1984), pp. 79-96.

Van den Broeck, Jan. *J. B. C. Verlooy: Vooruitstrevend Jurist en Politicus uit de 18de Eeuw*. Amsterdam: Standaard Wetenschappelij, 1980.

Van de Perre, A. *The Language Question in Belgium*. London: Richards, 1919.

Van der Essen, Léon. *A Short History of Belgium*. Chicago: University of Chicago Press, 1915.

Van der Haegen, Herman. "La nouvelle subdivision administrative en Belgique à la suite des recentes lois linguistiques." *Bulletin de la Société Belges d'Etudes Géographiques*, 33, 1 (1964), pp. 175-185.

Van der Haegen, Herman. "Spatial Pattern of the Evolution of the Population 1977-1981." *Bulletin de la Société Belge d'Etudes Géographiques*, 52, 2 (1982), pp. 248-253.

_____. *Atlas statistique du recensement de la population et du logement. Partie 1: données démographiques.* Brussels: Institut National de Statistique, 1983.

Van der Haegen, Herman; Pattyn, Martine; and Cardyn, C. "The Belgian Settlement System." *Acta Geographica Lovaniensia*, 22 (1982), pp. 251-363.

Vanderkindere, Léon. *La formation territoriale des principautés belges au Moyen Age.* 2 vols. Brussels: H. Lamertin, 1902.

Van Duinkerken, Anton. "Two Ways of Looking at Flanders: A Dutch View." *Delta*, 6, 4 (Winter 1963-1964), pp. 39-42.

Van Haegendoren, Maurits. *De Vlaamse Beweging Nu en Morgen.* 2 vols. Hasselt: Heideland, 1962.

_____. *The Flemish Movement in Belgium.* Translated from the Dutch. Antwerp: Flemish Cultural Council, 1965.

_____. "The Origins of the Language Shift in Flanders." *La Monda Lingvo-Problemo*, 1, 1 (January 1969), pp. 31-36.

_____. "Belgium and Its Double Language Boundary." *La Monda Lingvo-Problemo*, 2, 4 (January 1970), pp. 17-20.

Van Hecke, Etienne. "Finances et fiscalités communales: analyse cartographique." *Courrier Hebdomadaire du C.R.I.S.P.*, 1017/1018 (November 25, 1983).

Van Loey, A. *La langue néerlandaise en pays flamand.* Brussels: Office de Publicité, 1945.

Van Velthoven, Harry. "De Taalwetgeving en het Probleem Brussel, 1880-1914." *Taal en Sociale Integratie*, 4 (1981), pp. 247-259.

Vanwynsberghe, D. "Causes et conséquences macro-économiques de la régionalization belge." *Reflets et Perspectives de la Vie Economique*, 18, 4-5 (1979), pp. 283-296.

Velimsky, Vitezslav. "Belgium of the Eighties: Unitary, Bi-Cultural or Made Up of Three Regions?" *Europa Ethnica*, 40, 1 (1983), pp. 1-14.

Verdoodt, Albert. "Les problèmes communautaires belges à la lumière des études d'opinion." *Courrier Hebdomadaire du C.R.I.S.P.*, 742 (November 12, 1976).

_____. *Les problèmes des groupes linguistiques en Belgique.* Bibliothèque des Cahiers de l'Institut de Linguistique de Louvain, 10. Louvain: Editions Peeters, 1977.

_____. *Linguistic Tensions in Canadian and Belgian Labor Unions.* Quebec: International Center for Research on Bilingualism, 1977.

_____. "Introduction." Special issue on Belgium, *International Journal of the Sociology of Language*, 15 (1978), pp. 5-8.

_____. "Dix ans de recherches bibliographiques sur les problèmes communautaires belges." *Recherches Sociologiques*, 11, 2 (1980), pp. 237-245.

_____. *Bibliographie sur le problème linguistique belge.* Quebec: International Center for Research on Bilingualism, 1983.

Verhaegen, P. *La lutte scolaire en Belgique.* Ghent: Siffer, 1906.

Verlinden, Charles. *Les origines de la frontière linguistique en Belgique et la colonisation franque.* Brussels: La Renaissance du Livre, 1955.

Vermeylen, August. *Quelques aspects de la question des langues en Belgique.* Brussels: *Le Peuple*, 1919.

Vilrokx, Jacques. "The Flemish Elite in an Embattled Situation." *International Journal of the Sociology of Language*, 15 (1978), pp. 61-70.

De Vlaamsche Hoogeschool te Gent, 1916-1918. Ghent: Plantijn, 1919.

Vroede, M. de. *The Flemish Movement in Belgium.* Translated from the Dutch by W. Sanders. Antwerp: Kulturraad voor Vlaanderen & Institut voor Voorlichting, 1975.

"Wallonie-Bruxelles: de nouvelles réalités institutionnelles." *Wallonie Bruxelles*, 1 (October 1983), pp. 2-5.

Wetenschappelijk Onderzoek van de Brusselse Taaltoestanden. Vol. 1. Brussels: Nederlandse Commissie voor de Cultuur van de Brusselse Agglomeratie, 1974-1975.

Wigny. Pierre. *La troisième révision de la constitution*. Brussels: Bruylant, 1972.

Willemsen, A. W. *Het Vlaams-Nationalisme: De Geschiedenis van de Jaren 1914-1940*. 2nd ed. Utrecht: Ambo, 1969.

Wilmars, Dirk. *De Psychologie van de Franstalige in Vlaanderen*. Antwerp: Standaard-Boekhandel, 1966.

_____. *Le problème belge: la minorité francophone en Flandre*. Translated from the Dutch by Edmond Knaeps. Antwerp: Scriptoria, 1968.

Wilwerth, Claude. *Le statut linguistique de la fonction publique belge*. Brussels: Editions de l'Université de Bruxelles, 1980.

Witte, Els et al. *Histoire de Flandre des origines à nos jours*. Brussels: La Renaissance du Livre, 1983.

Zolberg, Aristide R. "The Making of Flemings and Walloons: Belgium: 1830-1914." *Journal of Interdisciplinary History*, 5, 2 (Fall 1974), pp. 179-235.

_____. "Transformation of Linguistic Ideologies: The Belgian Case." *Les états multilingues: problèmes et solutions*. Edited by Jean-Guy Savard and Richard Vigneault. Quebec: Les Presses de l'Université Laval, 1975. Pp. 445-472.

_____. "Les origines du clivage communautaire en Belgique," *Recherches Sociologiques*, 2 (October 1976), pp. 150-170.

_____. "Splitting the Difference: Federalization without Federalism in Belgium." *Ethnic Conflict in the Western World*. Edited by Milton J. Esman. Ithaca: Cornell University Press, 1977. Pp. 103-142.

_____. "Belgium." *Crises of Political Development in Europe and the United States*. Edited by Raymond Grew. Princeton: Princeton University Press, 1978. Pp. 99-138.

Unpublished Dissertations, Theses, Bibliographies and Papers

Callebaut, Inge Itta. "The Evolution of Language and Culture: Fifty Families in Brussels (1870-1986)." Paper presented at the Soviet-Belgian Symposium on Multilingualism: "Aspects of Interpersonal and Intergroup Communication in Pluricultural Societies," Bibliothèque Royale Albert 1er, Brussels, March 13-15, 1986.

Charlier, Jacques. "Les flux téléphoniques interzonaux en Belgique en 1982: une approche multivariée." Paper presented to the Study Group on the Geography of Communication, International Geographical Union, Montpelier, France, November 18-19, 1985.

Curtis, Arthur E. "New Perspectives on the History of the Language Problem in Belgium." Ph.D. dissertation, University of Oregon, 1971.

Dunn, James A., Jr. "Social Cleavage, Party Systems and Political Integration: A Comparison of the Belgian and Swiss Experiences." Ph.D. dissertation, University of Pennsylvania, 1970.

Mullier, Jean-Luc. "La géographie des mariages et les modèles gravitaires: le cas de Mouscron." Mémoire de Licencié en Géographie, Université Catholique de Louvain, 1980-1981.

Osinski, Lawrence. "Multi-Ethnicism and the Nation-State in Southeast Asia." M.A. thesis in the Department of Geography, University of Chicago, 1971.

Rouquette, Rémi. "Plurilinguisme et institutions politiques (Belgique, Canada, Luxembourg, Suisse)." DEA de Droit Publique, Paris X Nanterre, 1982.

Suenaert, Lieve, and Verdoodt, Pierre. "Langue et société en Belgique 1980-1985: Bibliographie analytique et guide du chercheur/Taal en Maatschappij Belgie 1980-1985: Analytische Bibliografie en Gids voor de Gebruiker." Centrum voor de Studie van de Pluriculturelle Maatschappij/Centre pour l'Etude de la Société Pluriculturelle, 1986.

Zoller, Henry. "Les différences objectives entre Flamands et Wallons." Mémoire de Licencié en Sciences Politiques et Sociales, Université Catholique de Louvain, 1963.

INDEX

Activists, 102-108, 113
Agnew, J., 23-24
agriculture, 80-82
airlines, 168
algemeen beschaafd Nederlands, 97
Amsterdam, 53
Antwerp, city of, 63, 65, 69, 71, 74, 75, 112, 154, 174, 183
Antwerp, province of, 71, 110
arbitration, 146, 148
armed forces, language in, 67, 92, 94, 105, 159; organization of, 108, 183
assimilation, 24
Austria, 45-47
autonomy, 141, 178; Flemish, 102, 104-106, 108, 113-115; Walloon, 100-101, 115, 127

Baetens Beardsmore, H., 33-34, 155
banks, 168
bar, Belgian, 169
Barth, F., 15
Battle of the Golden Spurs, 63, 175
Baudouin, King, 126, 181-182
Borms, A., 107-108, 112, 114-115
Bowman, I., 28
Brabant, medieval, 42-43
Brabant, province of, 71, 80, 87, 144, 181
Brel, J., 196
Brussels, city of, 52, 73, 75, 94, 103, 116-117, 119-120, 128, 134-138, 141-151, 154, 157, 159, 162-166, 174-179, 182, 184, 193-195; language in, 3-4, 46, 67, 78, 99, 114, 141; Middle Ages, 43, 45, see also communes, Brussels
Brussels, district of, 71, 110
Brussels Council, 143
Bruges, 43
Burgundy, Dukes of, 43

Cardyn, C., 163
Catholic clergy, 50
Catholic dioceses, 47
Catholic influence, 53, 82-88
Catholic political organizations, see Christian political organizations
Catholic schools, 72, 125-127
Catholic University of Louvain, 67, 93, 109, 112, 118, 139, 143, 165, 169
Cavin, J., 3
census, see language, census of
Charleroi, 50, 160
Charles V, 45
Charlier, J., 170
Christian political organizations, 61, 68-69, 74, 109, 127, 146, 148, 165-166
Claes, L., 127
Claeys, P., 37
class, social, 9, 21, 46, 48, 60, 69-70, 74, 79, 96, 97, 111, 131, 186
Clough, S., 95
Clout, H., 131
collaboration, 102, 107-108, 113, 120-121, 126
Comines, 134-135
Commission Flamande, 66-68
communes, Belgian, 53, 164, 172; around Brussels, 8, 96, 113, 116, 129, 132, 134, 136, 139, 144, 147-148, 154, 156-157, 160-162, 174; Brussels, 99, 117, 136, 156-157; Flemish, 73, 103; language border, 116-117, 120, 128-129, 132, 135-136, 154, 156-157, 161-162, 170; Walloon, 78, see also names of communes
Communist political organizations, 126-127
community councils, 143-152, 155, 164, 184-185
Concentration Wallonne, 119
Congo, Belgian, 131
Congress of Vienna, 48
Connor, W., 17-18
Conscience, H., 63-64
Conseil Economique Wallon, 167
consociational democracy, 36-37
Constitution, Belgian, 52, 58-61, 114, 140, 142-149, 151, 191-192
core-periphery model, 23-24, 26, 29
Coulon, M., 42
Council of Flanders, 104-105
Covell, M., 37
cultural councils, see community councils
culture, 13, 14-16, 23, 24, 28, 35, 38
Curtis, A., 60, 182

Declerck, H., 161
decrees, see German decrees; royal decrees
deficit, see public spending
de Laet, J., 69
De Lannoy, W., 161
de Maere, Baron, 69
demography, see population
De Raet, L., 92-93, 95
De Schryver, R., 77
Destrée, J., 101
Deutsch, K., 17, 20, 24
dialects, 3, 53, 60, 96-97, 134
Dikshit, R., 37-38
Dilbeek, 161
diplomatic service, language in, 67
discourse, the role of, 38
Dominion, L., 28
Dunkirk, 183
Dunn, J., 140, 192

East Flanders, province of, 61, 65, 71, 79, 82, 87, 110
economic councils, regional, 144, 147, 167
Economische Raad voor Vlaanderen, 167
education, language of, 46, 48, 63, 67, 92-94, 108, 130, 155-157; level of, 82-83; see also legislation, language
Egmont Pact, 147-148
Engheim, 173
England, 59
English, 146, 155-156
Esman, M., 23
ethnic, see ethnicity
ethnicity, the concept of, 14-18, 21, 22-24, 25-29, 32, 35, 36
European Community, 176

Fairchild, H., 27
fascism, 118-119
federalism, 77, 78, 101, 108, 112, 114, 115, 118-122, 127-128, 130, 132, 140, 144, 152
Flamenpolitik, 121
Flamingants, 66, 69, 74, 77-78, 93, 94, 99, 102, 110
Flanders, Francophones in, 8; medieval, 42-43; use of the term, 7, 9, 51-52, 58, 65, 67, 71, 77, 79, 89, 101-102
Flemish movement, rise of, 9-10, 47, 58, 63-70; 92-123, 127-128, 175, 190-191, 193
Fourons, 8, 134-135, 144, 168, 174, 181-182, 195
France, 47-48, 71, 120-121, 174
franchise, see suffrage
Frankish invasion, 42

Front Démocratique des Francophones, 139, 165
Front Party, 105-106, 108
Frontists, 105-108, 113

German decrees, 103-105, 121
German occupation, see World War I and World War II
German-speaking area, 3
Germanic languages, 1-3, 8-9, 41-43, 45, 49, 50, 51, 53, 134
Ghent, 43, 46, 63, 65, 75, 154-155, 174
Giddens, A., 30
Gladdish, K., 25-26
Gottmann, J., 27-28, 29
Grammens, F., 118
Gregory, D., 30

Hainaut, city of, 126
Hainaut, province of, 79, 87, 134-135
Halle-Vilvoorde, 136, 145
Happart, J., 181-182
Hapsburg Empire, 43, 45
Harmel Center, 129-130, 132-135
Hartshorne, R., 29
Hechter, M., 22-23
hinterland, see urban hinterlands

Industrial Revolution, 49-50, 92
industrialization, level of, 82, 130
information dichotomy, 170-173, 184
Institut Jules Destrée, 176
internal colonialism, 22, 29, 30
International Association of Workers, 70

Jennissen, E., 100
Jones, S., 29
judicial districts, 47, 130

Katholieke Vlaamsche Volkspartij,118
Kauppi, M., 21
Kluft and Jaspers survey, 161
Knight, D., 27-28, 29
Kohn, H., 17

landscape, linguistic, 173-174
language, census of, 4-5, 78, 116, 128-129, 132, 154, 157; geography of, 1, 3, 4, 53, 61, 112, 152, 154, 157; standardization of, 3, 46, 53, 96-97, 113; see also dialects; education, language of; Germanic languages; legislation, language; parliament, Belgian; Romance languages; royal decrees;

language boundary, demarcation of, 73, 118-119, 128, 130, 132-136, 139, 143-144, 181; function of, 173-174; location of, 41-45, 134
language shift, 98, 156
Latin, 42, 46, 48
law, the role of, 39-40; see also legislation
legislation, language, 39-40, 98, 113, 123, 128-130, 132, 135, 140-142, 154-155, 161, 165, 168, 190-191, 193-194; armed forces, 59, 73, 94, 111, 119; commercial establishments, 136-138, 146, 155; courts, 70-71, 73-74, 99, 100, 103-104, 119, 136, 154; education, 59, 71-72, 93-94, 98-99, 102, 116-117, 135-136; government and administration, 59, 65, 71-72, 74, 102, 108, 110-111, 113-117, 135-136, 154; period of Dutch rule, 48-49, 53; period of French rule, 48; see also Constitution, Belgian; German decrees; royal decrees
Leopold I, King, 59
Leopold III, King, 125-126, 182
Leuven, city of, 43, 112, 139, 154
Leuven, district of, 71, 110
Leven, H., 101
Levy, P., 157
Liberal political organizations, 61, 68-69, 109, 126-127, 146, 148, 165-166
liberalism, 21, 36, 58, 113, 117
libraries, 168-169
Liège, city of, 50, 73, 75, 76, 100, 115, 126, 163, 174
Liège, prince-bishopric of, 42, 43, 45, 47
Liège, province of, 61, 79, 86, 134-135
Lier, 48, 65
Ligue d'Action Wallonne, 112
Ligue Wallonne d'Ixelles, 76
Lijphart, A., 20, 24, 36-37
Lille, 174
Limburg, province of, 61, 70, 79, 87, 97, 105, 110, 135
Locke, J., 36
Lorwin, V., 117
Louis-Philippe, King, 59
Louise of Orléans, 59
Louvain, see Leuven, city of
Louvain-la-Neuve, 169
Luethy, H., 34
Luxembourg, Grand Duchy of, 45, 47, 48
Luxembourg, province of, 86, 127

McRae, K., 35, 151, 169, 179, 193
Mandel, E., 57
Markusen, A., 32
Martens, Prime Minister, 166, 181-182

Marxism, 21, 22, 79
Maximalists, 108, 110, 113
Mechelen, 174
migration, 45, 158-161
military, see armed forces
military districts, 61
Mill, J., 36
Minimalists, 108-110
modernization, 20, 22, 24
monarchy, Belgian, 58-59, 107, 125-126, 176
Moniteur Belge, 67, 74
Mons, 50, 174
Mouscron, 134-135
Mouvement Populaire Wallon, 132
Musée de la Vie Wallonne, 175

Nairn, T., 22-23
Namur, city of, 50, 104, 160, 162
Namur, province of, 86, 126
Napolean, 48, 59
nation, 17-18
nation-state, 13, 20, 28, 31
National Congress of 1830, 52
National Walloon Congress, 121
nationalism, state, 14, 17-19, 25-27, 30, 32-33, 51; in Belgium, 45, 51-52, 53, 68; see also substate nationalism
NATO, 176
Netherlands, the, 43, 45, 48-49, 50, 59, 114, 118
newspapers, Belgian, 172, 182
Nieuw Vlaanderen, 118
Nigeria, 192
Nivelles, 65
Nuttin, J., 178

opera, national, 169
organic-statism, 21

Pacifists, 102, 106-107
Palley, C., 25-26
Paris, 53
Parliament, Belgian, 64, 66, 69, 77, 92-93, 108, 112, 115, 129, 132, 135-136, 141, 143, 145-146, 149-150, 161; language use in, 74
partitions, administrative, 10, 95, 104, 106, 109-110, 113, 118, 132-136, 190; perceptual, 53
Pattyn, M., 163
personality, the concept of, 35, 37
Philip II, 45
physiographic regions, 53-54
Pirenne, H., 52
Platel, M., 78

Plavsic, W., 164
Polasky, J., 60
population, language regions, 5-6, 109, 114, 121, 141, 158-162, 184
positive rights, 39
Pred, A., 30
press, Belgian, 192
provinces, Belgian, 7, 47, 50, 52, 61, 64-65, 67, 69, 78, 80-88, 110, 164-165; see also names of provinces
public spending, 121, 168, 183-184, 195

Quebec, 184

radio, Belgian, 169, 180
Ragin, C., 25
railroads, Belgian, 68, 183
Rassemblement Wallon, 165
Rawkins, P., 23
regional councils, 143-144, 148-152, 164, 184-185
regionalism, the concept of, 27, 30-34, 35, 38, 189, 192-194
religion, geography of, 80, 82, 84-87, 186
Renard, A., 131-132
Retour à Liège party, 181
Revolution, Belgian, 6, 7, 49-51, 52, 63
Revue Wallonne, 76
Roman Empire, 41
Romance languages, 1-3, 8-9, 41-42, 49, 51, 53
Rotterdam, 183
Rousseau, J., 36
royal decrees, language academies, 72-73; language of eduction, 64, 65; language of laws, 111
royal question, 125-127, 140, 182
Ruys, M., 63

Sack, R., 31-32, 194
Sambre-Meuse Valley, 9, 50, 52, 60-61, 78, 80, 82, 97
school wars, 125-128, 130, 182
Schreurs, F., 78, 99, 121
Schreurs-Couvreur bill, 128
Seventeen Provinces, 43, 45
Sint-Genesius-Rode, 162, 179
Smith, A., 17, 25, 29
social theory, 30, 32, 153, 194
Socialist political organizations, 61, 109, 126-127, 146, 148, 165-166
Société Liègeoise de Littérature Wallonne, 75
Soja, E., 29, 30, 31-32
Spain, 45, 192
Sri Lanka, 192

state, 17, 24, 29, 36; see also West European state
Stengers, J., 42
stereotypes, 179-180
substate nationalism, the concept of, 19-28, 30, 32-34, 194
suffrage, 59, 74, 77, 97-98, 109, 113, 190
Swing, E., 116
Switzerland, ethnolinguistic tensions in, 33-34, 36, 53, 179, 191-192
Swyngedouw, E., 167
Symmons-Symonolewicz, K., 18, 27

taxes, 195
telephones, Belgian, 170
television, Belgian, 169, 180
territoriality, the concept of, 26-34, 36, 37-38, 58, 191-194
thiois, 43
toponoyms, 7

uneven development, 22, 29, 30
unions, trade, 131, 169-170, 186
University of Brussels, 67, 70, 169
University of Ghent, 64, 67, 70-72, 92-93, 95, 100, 104, 107-112
University of Louvain, see Catholic University of Louvain
urban heirarchy, 162-164
urban hinterlands, 163-164

Van Cauwelaert, 109, 114
Van der Haegen, H., 163
Van Haegendoren, M., 179
van Maerlant, J., 43
Van Nu en Straks, 94
Velimsky, V., 195
Verdoodt, A., 38, 135-136, 185
Verlooy, J., 46-47
Vlaamsch Nationaal Verbond, 118
Vlaamsch Studiecomité voor Politieke, Sociale en Cultureele Aangelegenheden, 114
Vlaanderen, 94
Volksunie, 127-128, 165, 177
Vos, H., 114-115

Wallonia, use of the term, 7, 9, 51-52, 58, 76, 77-79, 89, 99, 101-102, 122
Wallonie, 175
Walloon movement, rise of, 75-77, 99-101, 111-112, 114-115, 117, 119-123, 127, 131-132, 190-191
Waterloo, 48

West European states, 19-20
West Flanders, province of, 61, 65, 71, 79, 82, 87, 105, 110, 134
Whebell, C., 29
William I of Orange-Nassau, 41, 48-49, 50, 59-60
Williams, C., 29
World War I, 10, 28, 77-78, 92, 94, 95, 97, 99, 101, 102-107, 111-113, 120-121, 132, 190
World War II, 119-123, 125-129, 140-141, 155, 167, 181-182

Zeebrugge, 183

THE UNIVERSITY OF CHICAGO
GEOGRAPHY RESEARCH PAPERS
(Lithographed, 6 x 9 inches)

Titles in Print

48. BOXER, BARUCH. *Israeli Shipping and Foreign Trade.* 1957. x + 162 p.
56. MURPHY, FRANCIS C. *Regulating Flood-Plain Development.* 1958. x + 204 p.
62. GINSBURG, NORTON, ed. *Essays on Geography and Economic Development.* 1960. xx + 173 p.
71. GILBERT, EDMUND W. *The University Town in England and West Germany.* 1961. viii + 74 p.
72. BOXER, BARUCH. *Ocean Shipping in the Evolution of Hong Kong.* 1961. x + 95 p.
91. HILL, A. DAVID. *The Changing Landscape of a Mexican Municipio, Villa Las Rosas, Chiapas.* 1964. xiii + 121 p.
97. BOWDEN, LEONARD W. *Diffusion of the Decision To Irrigate: Simulation of the Spread of a New Resource Management Practice in the Colorado Northern High Plains.* 1965. xxvii + 146 p.
98. KATES, ROBERT W. *Industrial Flood Losses: Damage Estimation in the Lehigh Valley.* 1965. xi + 76 p.
101. RAY, D. MICHAEL. *Market Potential and Economic Shadow: A Quantitative Analysis of Industrial Location in Southern Ontario.* 1965. xvii + 164 p.
102. AHMAD, QAZI. *Indian Cities: Characteristics and Correlates.* 1965. viii + 184 p.
103. BARNUM, H. GARDINER. *Market Centers and Hinterlands in Baden-Württemberg.* 1966. xviii + 172 p.
105. SEWELL, W.R. DERRICK, ed. *Human Dimensions of Weather Modification.* 1966. xii + 423 p.
107. SOLZMAN, DAVID M. *Waterway Industrial Sites: A Chicago Case Study.* 1967. x + 138 p.
108. KASPERSON, ROGER E. *The Dodecanese: Diversity and Unity in Island Politics.* 1967. xiv + 184 p.
109. LOWENTHAL, DAVID, ed. *Environmental Perception and Behavior.* 1967. vi + 88 p.
112. BOURNE, LARRY S. *Private Redevelopment of the Central City, Spatial Processes of Structural Change in the City of Toronto.* 1967. xii + 199 p.
113. BRUSH, JOHN E., AND HOWARD L. GAUTHIER, JR. *Service Centers and Consumer Trips: Studies on the Philadelphia Metropolitan Fringe.* 1968. x + 182 p.
114. CLARKSON, JAMES D. *The Cultural Ecology of a Chinese Village: Cameron Highlands, Malaysia.* 1968. xiv + 174 p.
115. BURTON, IAN, ROBERT W. KATES, AND RODMAN E. SNEAD. *The Human Ecology of Coastal Flood Hazard in Megalopolis.* 1968. xiv + 196 p.
117. WONG, SHUE TUCK. *Perception of Choice and Factors Affecting Industrial Water Supply Decisions in Northeastern Illinois.* 1968. x + 93 p.
118. JOHNSON, DOUGLAS L. *The Nature of Nomadism: A Comparative Study of Pastoral Migrations in Southwestern Asia and Northern Africa.* 1969. viii + 200 p.

119. DIENES, LESLIE. *Locational Factors and Locational Developments in the Soviet Chemical Industry.* 1969. x + 262 p.
120. MIHELIC, DUESAN. *The Political Element in the Port Geography of Trieste.* 1969. ix + 104 p.
121. BAUMANN, DUANE D. *The Recreational Use of Domestic Water Supply Reservoirs: Perception and Choice.* 1969. ix + 125 p.
122. LIND, AULIS O. *Coastal Landforms of Cat Island, Bahamas: A Study of Holocene Accretionary Topography and Sea-Level Change.* 1969. ix + 156 p.
123. WHITNEY, JOSEPH B. R. *China: Area, Administration and Nation Building.* 1970. xiii + 198 p.
124. EARICKSON, ROBERT. *The Spatial Behavior of Hospital Patients: A Behavioral Approach to Spatial Interaction in Metropolitan Chicago.* 1970. xi + 138 p.
125. DAY, JOHN C. *Managing the Lower Rio Grande: An Experience in International River Development.* 1970. xii + 274 p.
126. MacIVER, IAN. *Urban Water Supply Alternatives: Perception and Choice in the Grand Basin, Ontario.* 1970. ix + 178 p.
127. GOHEEN, PETER G. *Victorian Toronto, 1850 to 1900: Pattern and Process of Growth.* 1970. xiii + 278 p.
128. GOOD, CHARLES M. *Rural Markets and Trade in East Africa.* 1970. xvi + 252 p.
129. MEYER, DAVID R. *Spatial Variation of Black Urban Households.* 1970. xiv + 127 p.
130. GLADFELTER, BRUCE G. *Meseta and Campina Landforms in Central Spain: A Geomorphology of the Alto Henares Basin.* 1971. xii + 204 p.
131. NEILS, ELAINE M. *Reservation to City: Indian Migration and Federal Relocation.* 1971. x + 198 p.
132. MOLINE, NORMAN T. *Mobility and the Small Town, 1900-1930.* 1971. ix + 169 p.
133. SCHWIND, PAUL J. *Migration and Regional Development in the United States, 1950-1960.* 1971. x + 170 p.
134. PYLE, GERALD F. *Heart Disease, Cancer and Stroke in Chicago: A Geographical Analysis with Facilities, Plans for 1980.* 1971. ix + 292 p.
135. JOHNSON, JAMES F. *Renovated Waste Water: An Alternative Source of Municipal Water Supply in the United States.* 1971. ix + 155 p.
136. BUTZER, KARL W. *Recent History of an Ethiopian Delta: The Omo River and the Level of Lake Rudolf.* 1971. xvi + 184 p.
139. McMANIS, DOUGLAS R. *European Impressions of the New England Coast, 1497-1620.* 1972. viii + 147 p.
140. COHEN, YEHOSHUA S. *Diffusion of an Innovation in an Urban System: The Spread of Planned Regional Shopping Centers in the United States, 1949-1968.* 1972. ix + 136 p.
141. MITCHELL, NORA. *The Indian Hill-Station: Kodaikanal.* 1972. xii + 199 p.
142. PLATT, RUTHERFORD H. *The Open Space Decision Process: Spatial Allocation of Costs and Benefits.* 1972. xi + 189 p.
143. GOLANT, STEPHEN M. *The Residential Location and Spatial Behavior of the Elderly: A Canadian Example.* 1972. xv + 226 p.
144. PANNELL, CLIFTON W. *T'ai-Chung, T'ai-wan: Structure and Function.* 1973. xii + 200 p.

145. LANKFORD, PHILIP M. *Regional Incomes in the United States, 1929-1967: Level, Distribution, Stability, and Growth.* 1972. x + 137 p.
146. FREEMAN, DONALD B. *International Trade, Migration, and Capital Flows: A Quantitative Analysis of Spatial Economic Interaction.* 1973. xiv + 201 p.
147. MYERS, SARAH K. *Language Shift Among Migrants to Lima, Peru.* 1973. xiii + 203 p.
148. JOHNSON, DOUGLAS L. *Jabal al-Akhdar, Cyrenaica: An Historical Geography of Settlement and Livelihood.* 1973. xii + 240 p.
149. YEUNG, YUE-MAN. *National Development Policy and Urban Transformation in Singapore: A Study of Public Housing and the Marketing System.* 1973. x + 204 p.
150. HALL, FRED L. *Location Criteria for High Schools: Student Transportation and Racial Integration.* 1973. xii + 156 p.
151. ROSENBERG, TERRY J. *Residence, Employment, and Mobility of Puerto Ricans in New York City.* 1974. xi + 230 p.
152. MIKESELL, MARVIN W., ed. *Geographers Abroad: Essays on the Problems and Prospects of Research in Foreign Areas.* 1973. ix + 296 p.
153. OSBORN, JAMES. *Area, Development Policy, and the Middle City in Malaysia.* 1974. x+ 291 p.
154. WACHT, WALTER F. *The Domestic Air Transportation Network of the United States.* 1974. ix + 98 p.
155. BERRY, BRIAN J. L. et al. *Land Use, Urban Form and Environmental Quality.* 1974. xxiii + 440 p.
156. MITCHELL, JAMES K. *Community Response to Coastal Erosion: Individual and Collective Adjustments to Hazard on the Atlantic Shore.* 1974. xii + 209 p.
157. COOK, GILLIAN P. *Spatial Dynamics of Business Growth in the Witwatersrand.* 1975. x + 144 p.
159. PYLE, GERALD F. et al. *The Spatial Dynamics of Crime.* 1974. x + 221 p.
160. MEYER, JUDITH W. *Diffusion of an American Montessori Education.* 1975. xi + 97 p.
161. SCHMID, JAMES A. *Urban Vegetation: A Review and Chicago Case Study.* 1975. xii + 266 p.
162. LAMB, RICHARD F. *Metropolitan Impacts on Rural America.* 1975. xii + 196 p.
163. FEDOR, THOMAS STANLEY. *Patterns of Urban Growth in the Russian Empire during the Nineteenth Century.* 1975. xxv + 245 p.
164. HARRIS, CHAUNCY D. *Guide to Geographical Bibliographies and Reference Works in Russian or on the Soviet Union.* 1975. xviii + 478 p.
165. JONES, DONALD W. *Migration and Urban Unemployment in Dualistic Economic Development.* 1975. x + 174 p.
166. BEDNARZ, ROBERT S. *The Effect of Air Pollution on Property Value in Chicago.* 1975. viii + 111 p.
167. HANNEMANN, MANFRED. *The Diffusion of the Reformation in Southwestern Germany, 1518-1534.* 1975. ix + 235 p.
168. SUBLETT, MICHAEL D. *Farmers on the Road. Interfarm Migration and the Farming of Noncontiguous Lands in Three Midwestern Townships. 1939-1969.* 1975. xiii + 214 p.

169. STETZER, DONALD FOSTER. *Special Districts in Cook County: Toward a Geography of Local Government.* 1975. xi + 177 p.
171. SPODEK, HOWARD. *Urban-Rural Integration in Regional Development: A Case study of Saurashtra, India—1800-1960.* 1976. xi + 144 p.
172. COHEN, YEHOSHUA S., AND BRIAN J. L. BERRY. *Spatial Components of Manufacturing Change.* 1975. vi + 262 p.
173. HAYES, CHARLES R. *The Dispersed City: The Case of Piedmont, North Carolina.* 1976. ix + 157 p.
174. CARGO, DOUGLAS B. *Solid Wastes: Factors Influencing Generation Rates.* 1977. viii + 100 p.
175. GILLARD, QUENTIN. *Incomes and Accessibility. Metropolitan Labor Force Partici-pation, Commuting, and Income Differentials in the United States, 1960-1970.* 1977. ix + 106 p.
176. MORGAN, DAVID J. *Patterns of Population Distribution: A Residential Preference Model and Its Dynamic.* 1978. xiii + 200 p.
177. STOKES, HOUSTON H., DONALD W. JONES, AND HUGH M. NEUBURGER. *Unemployment and Adjustment in the Labor Market: A Comparison between the Regional and National Responses.* 1975. ix + 125 p.
180. CARR, CLAUDIA J. *Pastoralism in Crisis. The Dasanetch and their Ethiopian Lands.* 1977. xx + 319 p.
181. GOODWIN, GARY C. *Cherokees in Transition: A Study of Changing Culture and Environment Prior to 1775.* 1977. ix + 207 p.
182. KNIGHT, DAVID B. *A Capital for Canada: Conflict and Compromise in the Nineteenth Century.* 1977. xvii + 341 p.
183. HAIGH, MARTIN J. *The Evolution of Slopes on Artificial Landforms, Blaenavon, U.K..* 1978. xiv + 293 p.
184. FINK, L. DEE. *Listening to the Learner: An Exploratory Study of Personal Meaning in College Geography Courses.* 1977. ix + 186 p.
185. HELGREN, DAVID M. *Rivers of Diamonds: An Alluvial History of the Lower Vaal Basin, South Africa.* 1979. xix + 389 p.
186. BUTZER, KARL W., ed. *Dimensions of Human Geography: Essays on Some Familiar and Neglected Themes.* 1978. vii + 190 p.
187. MITSUHASHI, SETSUKO. *Japanese Commodity Flows.* 1978. x + 172 p.
188. CARIS, SUSAN L. *Community Attitudes toward Pollution.* 1978. xii + 211 p.
189. REES, PHILIP M. *Residential Patterns in American Cities: 1960.* 1979. xvi + 405 p.
190. KANNE, EDWARD A. *Fresh Food for Nicosia.* 1979. x + 106 p.
192. KIRCHNER, JOHN A. *Sugar and Seasonal Labor Migration: The Case of Tucumán, Argentina.* 1980. xii + 174 p.
193. HARRIS, CHAUNCY D., AND JEROME D. FELLMANN. *International List of Geographical Serials Third Edition, 1980.* 1980. vi + 457 p.
194. HARRIS, CHAUNCY D. *Annotated World List of Selected Current Geographical Serials Fourth Edition. 1980.* 1980. iv + 165 p.
195. LEUNG, CHI-KEUNG. *China: Railway Patterns and National Goals.* 1980. xv + 243 p.
196. LEUNG, CHI-KEUNG, AND NORTON S. GINSBURG, eds. *China: Urbanizations and National Development.* 1980. ix + 283 p.

197. DAICHES, SOL. *People in Distress: A Geographical Perspective on Psychological Well-being.* 1981. xiv + 199 p.

198. JOHNSON, JOSEPH T. *Location and Trade Theory: Industrial Location, Comparative Advantage, and the Geographic Pattern of Production in the United States.* 1981. xi + 107 p.

199-200. STEVENSON, ARTHUR J. *The New York-Newark Air Freight System.* 1982. xvi + 440 p.

201. LICATE, JACK A. *Creation of a Mexican Landscape: Territorial Organization and Settlement in the Eastern Puebla Basin, 1520-1605.* 1981. x + 143 p.

202. RUDZITIS, GUNDARS. *Residential Location Determinants of the Older Population.* 1982. x + 117 p.

203. LIANG, ERNEST P. *China: Railways and Agricultural Development, 1875-1935.* 1982. xi + 186 p.

204. DAHMANN, DONALD C. *Locals and Cosmopolitans: Patterns of Spatial Mobility during the Transition from Youth to Early Adulthood.* 1982. xiii + 146 p.

205. FOOTE, KENNETH E. *Color in Public Spaces: Toward a Communication-Bases Theory of the Urban Built Environment.* 1983. xiv + 153 p.

206. HARRIS, CHAUNCY D. *Bibliography of Geography. Part II: Regional. Volume 1. The United States of America.* 1984. viii + 178. p.

207-208. WHEATLEY, PAUL. *Nagara and Commandery: Origins of the Southeast Asian Urban Traditions.* 1983. xv + 472 p.

209. SAARINEN, THOMAS F., DAVID SEAMON, AND JAMES L. SELL, eds. *Environmental Perception and Behavior: An Inventory and Prospect.* 1984. x + 263 p.

210. WESCOAT, JAMES L., JR. *Integrated Water Development: Water Use and Conservation Practice in Western Colorado.* 1984. xi + 239 p.

211. DEMKO, GEORGE J., AND ROLAND J. FUCHS, eds. *Geographical Studies on the Soviet Union: Essays in Honor of Chauncy D. Harris.* 1984. vii + 294 p.

212. HOLMES, ROLAND C. *Irrigation in Southern Peru: The Chili Basin.* 1986. ix + 199 p.

213. EDMONDS, RICHARD Louis. *Northern Frontiers of Qing China and Tokugawa Japan: A Comparative Study of Frontier Policy.* 1985. xi + 209 p.

214. FREEMAN, DONALD B., AND GLEN B. NORCLIFFE. *Rural Enterprise in Kenya: Development and Spatial Organization of the Nonfarm Sector.* 1985. xiv + 180 p.

215. COHEN, YEHOSHUA S., AND AMNON SHINAR. *Neighborhoods and Friendship Networks: A Study of Three Residential Neighborhoods in Jerusalem.* 1985. ix + 137 p.

217-218. CONZEN, MICHAEL P., ed. *World Patterns of modern Urban Change: Essays in Honor of Chauncy D. Harris.* 1986. x + 479.

219. KOMOGUCHI, YOSHIMI. *Agricultural Systems in the Tamil Nadu: A Case Study of Peruvalanallur Village.* 1986. xvi + 175 p.

220. GINSBURG, NORTON, JAMES OBORN, AND GRANT BLANK. *Geographic Perspectives on the Wealth of Nations.* 1986. ix + 1331 p.

221. BAYLSON, JOSHUA C. *Territorial Allocation by Imperial Rivalry: The Human Legacy in the Near East.* 1987. xi + 138 p.

224. PLATT, RUTHERFORD H., SHEILA G. PELCZARSKI, AND BARBARA K. BURBANK, eds. *Cities on the Beach: Management Issues of Developed Coastal Barriers.* 1987. vii + 324 p.
227. MURPHY, ALEXANDER B. *The Regional Dynamics of Language Differentiation in Belgium: A Study in Cultural-Political Geography.* 1988. xiii + 249 p.

Forthcoming Titles:

222. DORN, MARILYN A. *The Administrative Partitioning of Costa Rica: Politics and Planners.*
223. ASTROTH, JOSEPH H. JR. *Understanding Peasant Agriculture: An Integrated Land-Use Model for the Punjab.*
225. LATZ, G. IRVING III. *Agricultural Development in Japan: The Land Improvement District in Concept and Practice.*
226. GRITZNER, JEFFREY A. *The West African Sahel: Human Agency and Environmental Change.*